Wings over the Great Plains: Bird Migrations in the Central Flyway

Paul A. Johnsgard

Abstract

The Central Flyway has been recognized as a collective North-South migratory pathway centered on the North American Great Plains for nearly a century, but it has never been analyzed as the species that most closely follow it, or the major stopping points used by those species on their journeys between their northern breeding and southern wintering grounds. A total of 114 U.S. and 21 Canadian localities of special importance to birds migrating within the Central Flyway are identified and described in detail. Judging from available regional, state and local information, nearly 400 species of 50 avian families regularly use the Central Flyway during their migrations. Nearly 90 Central Flyway species have wintering areas partly extending variably far into the Neotropic zoogeographic realm, and at least 50 of these winter entirely within the Neotropic realm. A few of these species undertake some of the longest known migrations of all birds, in excess of 8.000 miles in each direction. Seven maps, 49 figures and over 100 literature citations are included.

Wings over the Great Plains:

Bird Migrations in the Central Flyway

Paul A. Johnsgard

University of Nebraska-Lincoln

Zea Books • Lincoln, Nebraska • 2012

ısʙɴ 978-1-60962-028-8 paperback
ısʙɴ 978-1-60962-029-5 ebook

Set in Palatino Linotype. Design and composition by Paul Royster.
Zea Books are published by the University of Nebraska–Lincoln Libraries.

Electronic (pdf) edition available online at http://digitalcommons.unl.edu/zeabook/
Print edition can be ordered from http://www.lulu.com/spotlight/unllib

Contents

List of Maps and Figures 7

1. Introduction to the North American Flyways 9

2. Important Refuges, Sanctuaries, and other
   Sites of Significance to Migratory Birds
   in the Central Plains 22

3. Geographic Distributions and Migration Patterns
   of Migratory Bird Species in the Central Plains 77

Appendix. Taxonomic List of Species Mentioned
   in the Text 227

Literature Cited & Selective Bibliography 239

List of Maps & Figures

All illustrations by Paul A. Johnsgard

Maps

1. Major climax plant formations of North America. 10–11
2. Principal breeding and wintering areas of North American
    waterfowl and administrative flyway units. 12–13
3. Native plant communities of the Great Plains 14
4. Approximate boundaries of the Central Flyway 16
5. Approximate boundaries of the Pacific, Mississippi
    and Atlantic flyways 17
6. Approximate boundaries of the Central Flyway
    south of Canada 19
7. Approximate boundaries of the Central Flyway in Canada . . . 20

Figures

1. Black-bellied Whistling Duck. 86
2. Greater White-fronted Goose. 88
3. Snow Goose. 89
4. Canada Goose 91
5. Tundra Swan 93
6. American Wigeon 95
7. Northern Pintail 99
8. Redhead pair 102
9. Long-tailed Duck. 105
10. American Bittern, Double-crested Cormorant
    and Pied-billed Grebe 108
11. Western Grebe 109
12. American White Pelican 111
13. Osprey. 117
14. Bald Eagle 119
15. Red-tailed Hawk 123

16. Ferruginous Hawk 125
17. Golden Eagle 126
18. Peregrine Falcon 129
19. Prairie Falcon 130
20. Whooping Crane 134
21. Willet. 138
22. Lesser Yellowlegs 139
23. Long-billed Curlew 140
24. Eskimo Curlew 141
25. Ruddy Turnstone 142
26. Sanderling 144
27. Short-billed Dowitcher 146
28. Wilson's Snipe 148
29. American Woodcock 149
30. Wilson's Phalarope 150
31. Common Tern. 154
32. Forster's Tern 155
33. White-winged Dove 156
34. Burrowing Owl 159
35. Saw-whet Owl 161
36. Common Nighthawk. 163
37. Ruby-throated Hummingbird. 165
38. Scissor-tailed Flycatcher 171
39. Red-eyed Vireo 174
40. Blue Jay 176
41. Horned Lark, American Crow and Sage Thrasher 178
42. White-breasted Nuthatch 182
43. Marsh Wren 184
44. Eastern Bluebird 187
45. European Starling 191
46. Black-and-White Warbler 196
47. Western Tanager 212
48. Red-winged Blackbird.. 215
49. Yellow-headed Blackbird 217

1

Introduction to the
North American Flyways

North America is a vast continent, with a variety of natural biological communities ranging from extreme desert to tropical rainforests (Map 1). Partly because of this biological diversity, including areas rich in food resources and wetlands, and partly because of topographic differences such as mountain ranges and river valleys, there are great differences in the ease with which animals can move about, especially from north to south and vice versa. These factors are especially important for water-dependent animals such as waterfowl, which usually undertake long migrations between the northern latitudes, where summers may be short but temperatures and food supplies are favorable for breeding, and more southern latitudes where temperatures allow for overwintering without danger of freezing to death. Over time, many species of waterfowl (ducks, geese and swans) evolved to exploit these important regional differences in seasonally available resources, and came to share geographically wide separated breeding and wintering areas (Map 2).

As a region rich in wetlands and sufficient amounts of precipitation to permit perennial grasslands to evolve and survive, the Great Plains of North America host the continent's richest and most extensive grasslands (Map 3), the preferred nesting cover for many species of birds, and the wetter parts of the great Plains are noted for their abundant marshy wetlands, the prime nesting areas for both waterfowl and shorebirds.

Comparing the map of the historic wintering and breeding areas of North American waterfowl, as of the 1960's (Map 2), with a map of the native vegetation of the Great Plains (Map 3), it may be seen that the Great Plains region has historically supported the continent's densest concentration of breeding ducks. This concentration is mainly centered

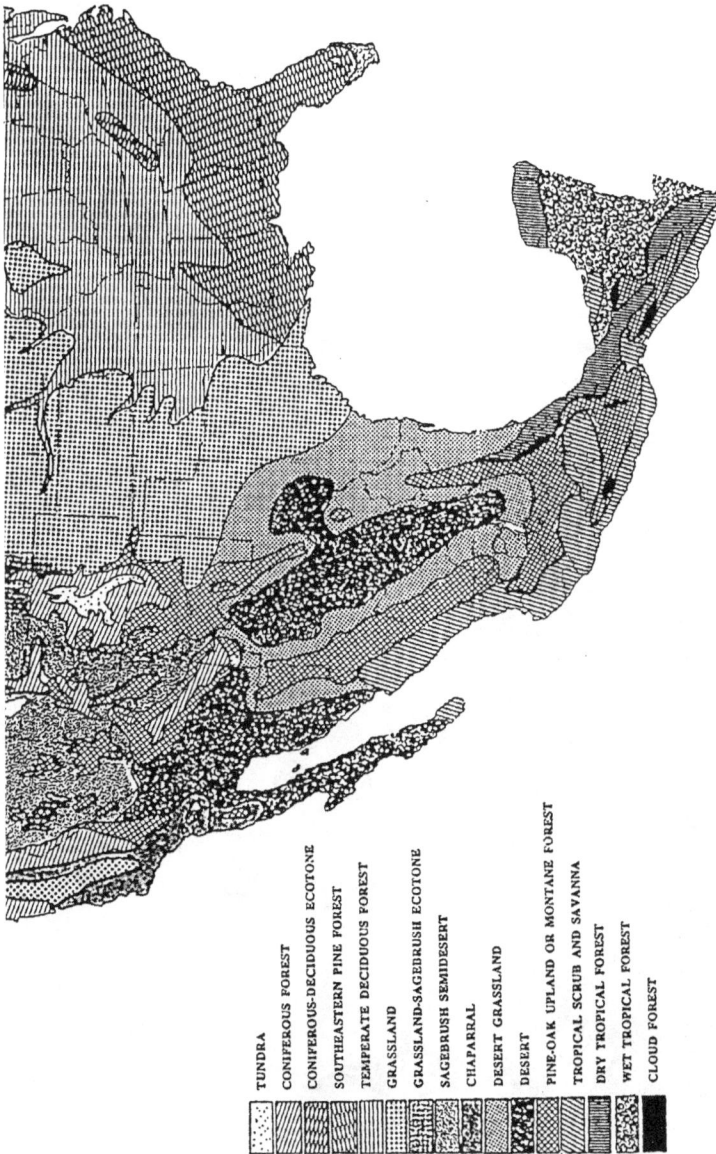

Map 1. Major climax plant formations of North America. Illustration by author, after various sources.

TUNDRA

CONIFEROUS FOREST

CONIFEROUS-DECIDUOUS ECOTONE

SOUTHEASTERN PINE FOREST

TEMPERATE DECIDUOUS FOREST

GRASSLAND

GRASSLAND-SAGEBRUSH ECOTONE

SAGEBRUSH SEMIDESERT

CHAPARRAL

DESERT GRASSLAND

DESERT

PINE-OAK UPLAND OR MONTANE FOREST

TROPICAL SCRUB AND SAVANNA

DRY TROPICAL FOREST

WET TROPICAL FOREST

CLOUD FOREST

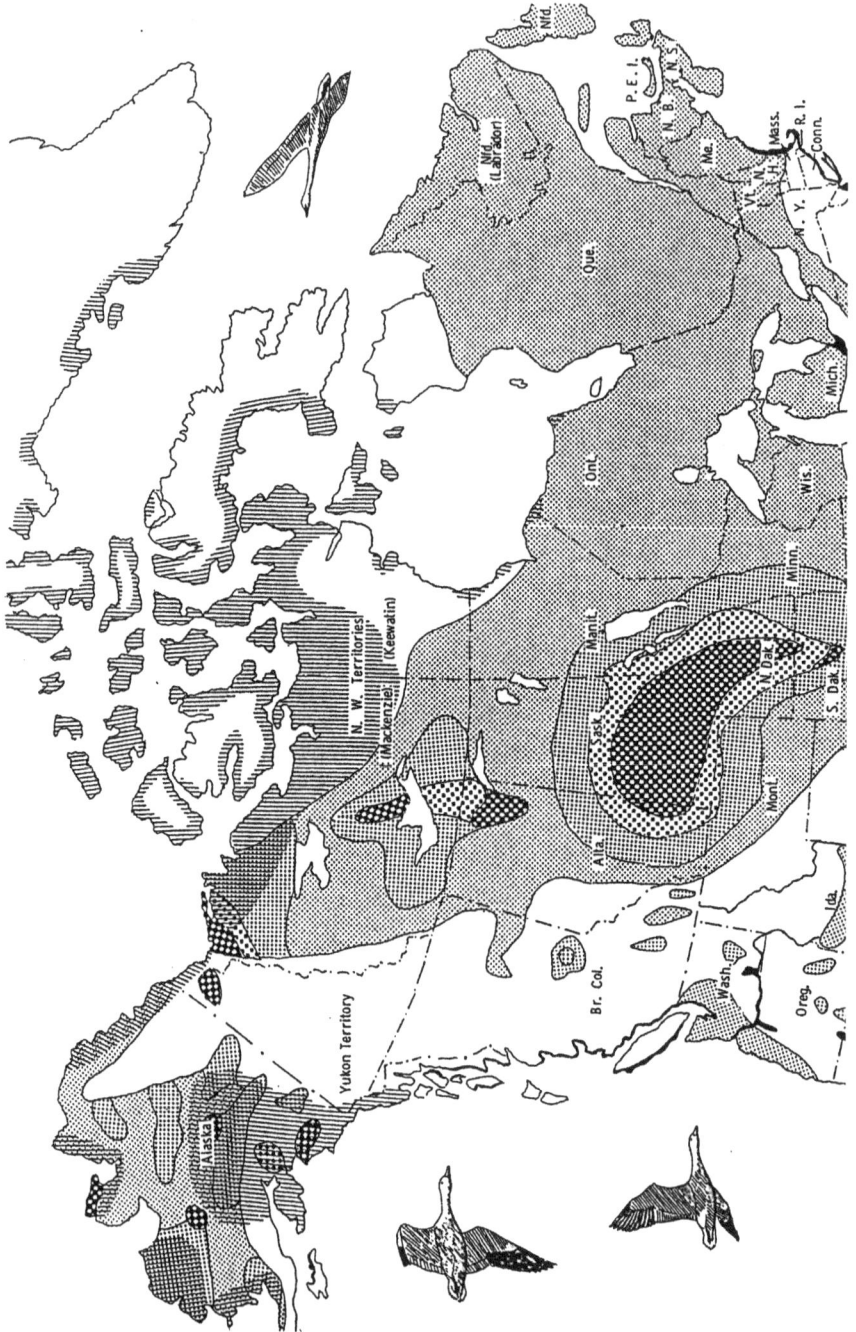

Map 2. Principal breeding and wintering areas of North American waterfowl and administrative flyway units. Breeding ground information based in part of Linduska (1964); wintering areas from various sources. Illustration by author.

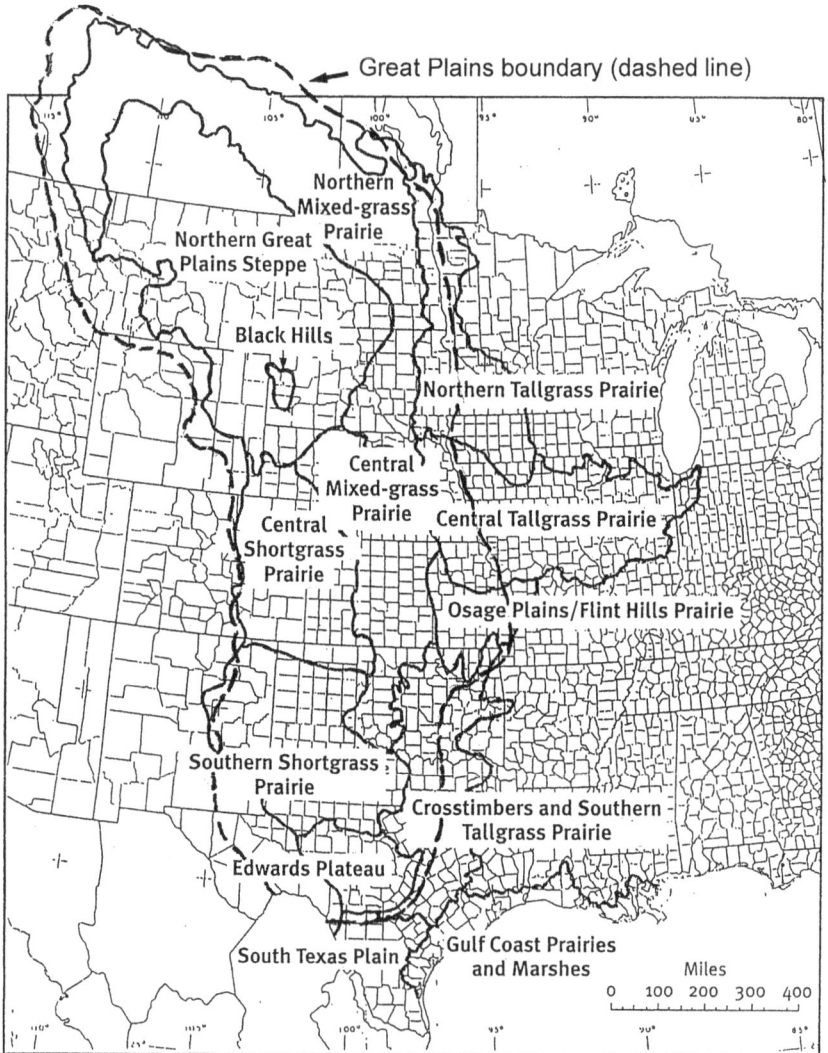

Map 3. Native plant communities of the Great Plains. Adapted by author from Johnsgard (2003); boundary of Great Plains after Wishart (2004).

in the "duck factory" region of perennial gasses and marshy wetlands of the Dakotas and the prairie provinces of southern Canada. From the Gulf coast of Texas, to the northern limits of the Great Plains in Canada, the region's uniform topography and abundance of wetlands provides a natural passageway for birds moving between wintering and breeding areas. Such passageways have been called "flyways,"

Somehow the word flyway seems somewhat magical; conjuring up the idea of invisible flightlines for birds to follow on their seasonal movements back and forth between their wintering and breeding grounds The word is well entrenched in older ornithological literature, but its current usage was formulated by Frederick C. Lincoln. For many years during the early part of the 20th century Lincoln was in charge of a program devoted to analyzing the distribution and migration of birds, in the Fish and Wildlife Service's Division of Wildlife Research. He stated (1935, 1943) that the terms "migration route" and "flyway' had historically been used rather indiscriminately. He therefore re-defined migration routes as those lanes of avian travel between an individual bird's winter and breeding quarters, whereas flyways were considered by him to be broader areas into which various individual migration routes blend, within a definite geographic region. Flyways were thus defined as rather broad arterial boulevards, to which migration routes of individual birds act as tributaries. A more modern definition might be: Flyways are latitudinal migration routes in North America that are used seasonally by many bird species, and can thus be mapped relative to broadly defined geographic features (Maps 4 & 5). Newton (2008) similarly defined a flyway as "An established air route used year after year by large numbers of migratory birds..."

Lincoln observed that distinct waterfowl flyways are most evident in southern parts of North America, from about 45° N. latitude south to the Gulf coast, since farther to the north the Canadian and Alaska arctic consists of one vast breeding ground that is used by many species of waterfowl. Lincoln geographically defined four American flyways on the basis of data from banded waterfowl, since banding return and recover data showed that these identifiable ducks and geese tended to follow individual ancestral routes, perpetuating not only the routes but also the groups of birds using them.

Because Lincoln's proposed flyways were based entirely on waterfowl movements, it is not surprising that they also apply well to the migrations of other water-dependent birds, such as loon, grebes, cranes, shorebirds, and similar relatively aquatic birds. They bear little or no relationship to the migrations of arboreal birds, including

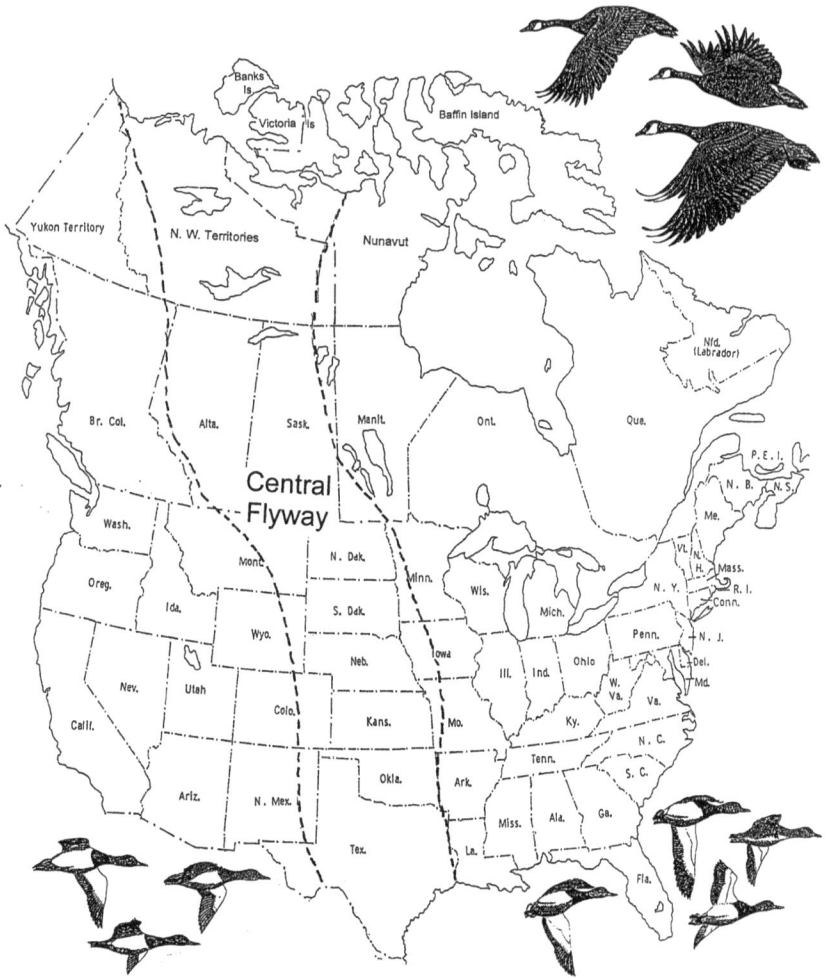

Map 4. Approximate boundaries of the Central Flyway. Illustration by author.

nearly all passerines. But, at least in the Central Flyway, a number of grassland passerine species (such as various sparrows and longspurs) have migration routes that closely correspond to the Central Flyway's configuration.

Soaring species such as many raptors often utilize topographic features that provide slope-effect winds or thermal updrafts and their

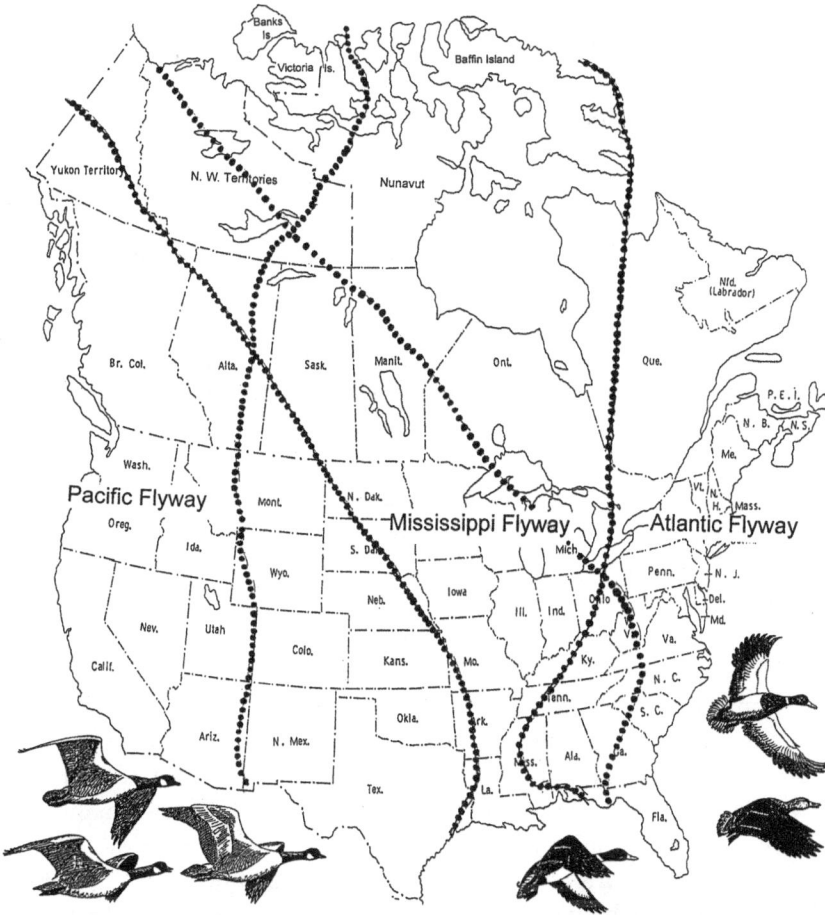

Map 5. Approximate boundaries of the Pacific, Mississippi and Atlantic flyways. For clarity, the limits of the Mississippi Flyway are shown by connected dots. Illustration by author.

movement often follow mountain ranges rather than wetland distributions or river courses, so their migratory relationships to traditional flyway boundaries are often negligible. Migratory routes of many bird species are also often influenced by long-term ecological conditions and climatic factors, and short-term changes in weather, individual experience, and local food availability.

It is now known that waterfowl, especially female ducks, have a high level of site-fidelity, and when mature tend to return to the area where they hatched and first acquired the ability to fly. In this way, migratory "traditions" gradually develop, with predictable short-term stopping points ("staging areas") and regularly used summer and winter end-points (Hochbaum, 1967). Sometimes such tradition-ally used locations become "burned out," owing to over-hunting or ecological changes, and it may be a long time before that area again becomes part of a species' migratory tradition.

In spite of such limitations to the flyway concept, it is well en-trenched in the popular literature, and has been adapted by the U.S. Fish and Wildlife Service and the Canadian Wildlife Service to define large-scale administrative boundaries (Map 2). The states adminis-tratively included in the Central Flyway include Texas, Oklahoma, Kansas, Nebraska, South Dakota, and North Dakota. In Canada, Al-berta, Saskatchewan, and the Northwest Territories are considered to fall within the boundaries of the Central Flyway. In terms of game management, hunting seasons and harvest limits are often defined ac-cording to such flyway entities. Given their associated geo-political and biological importance, the flyway concept is significant enough to warrant a thorough review.

As defined by Lincoln, the four North American flyways, from east to west are the Atlantic Flyway, the Mississippi Flyway, the Central Fly-way, and the Pacific Flyway (Maps 4 & 5). The Central Flyway is mapped in Map 4, with some modifications from Lincoln's original conceptual limits, based on currently known waterfowl migration routes. Specifi-cally, the western boundary of the Central Flyway is here considered to be the eastern base of the Rocky Mountains, and its eastern edges have been extended to include the mainly prairie-dominated western parts of Missouri, Iowa, and Minnesota, plus southwestern Manitoba. The other three flyways are shown in Map 5 and as noted above, their limits tend to blur and overlap at more northern latitudes.

The Central Flyway also generally falls within the geographic lim-its of North America's historic native grasslands, where small to large wetlands were historically abundant, and provided a myriad of avail-able resting and foraging areas for water-dependent migrants (Map 4). These native grasslands, especially those in the wetter, northern and eastern regions of the Great Plains, have now virtually vanished, except

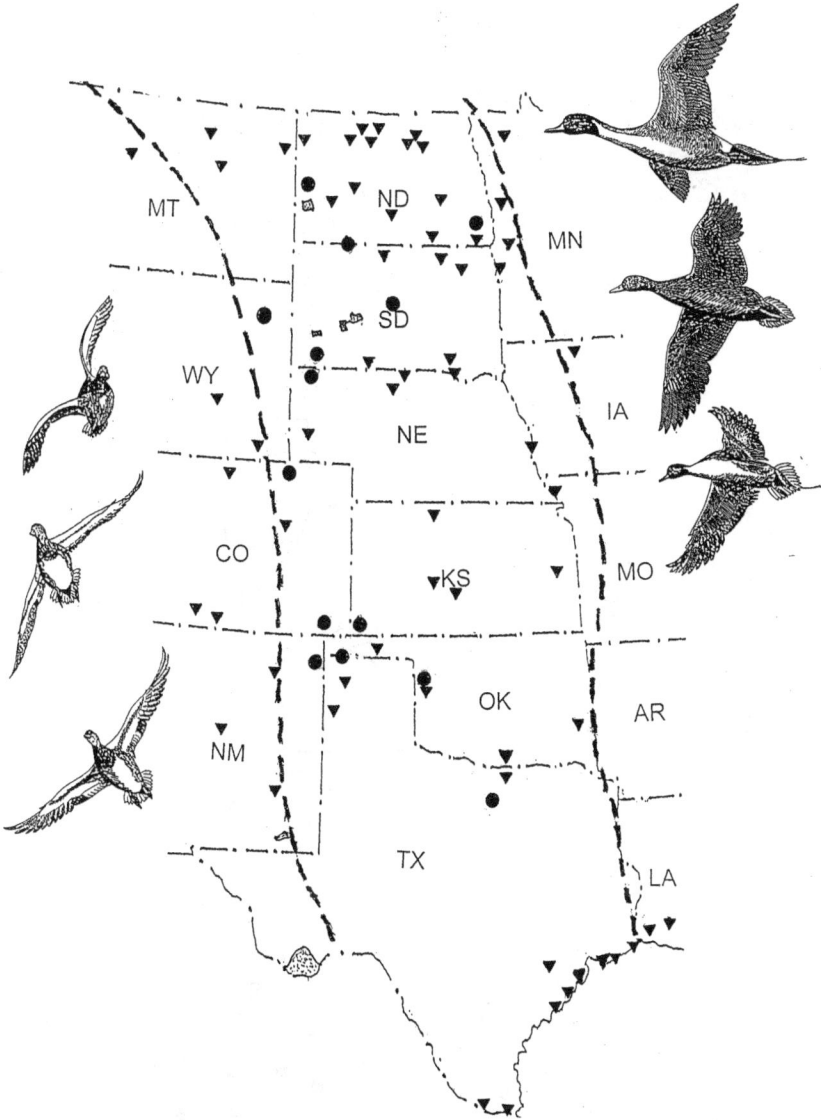

Map 6. Approximate boundaries of the Central Flyway south of Canada. National wildlife refuges are shown as triangles, national grasslands as circles, and national parks as enclosed and stippled areas. Illustration by author.

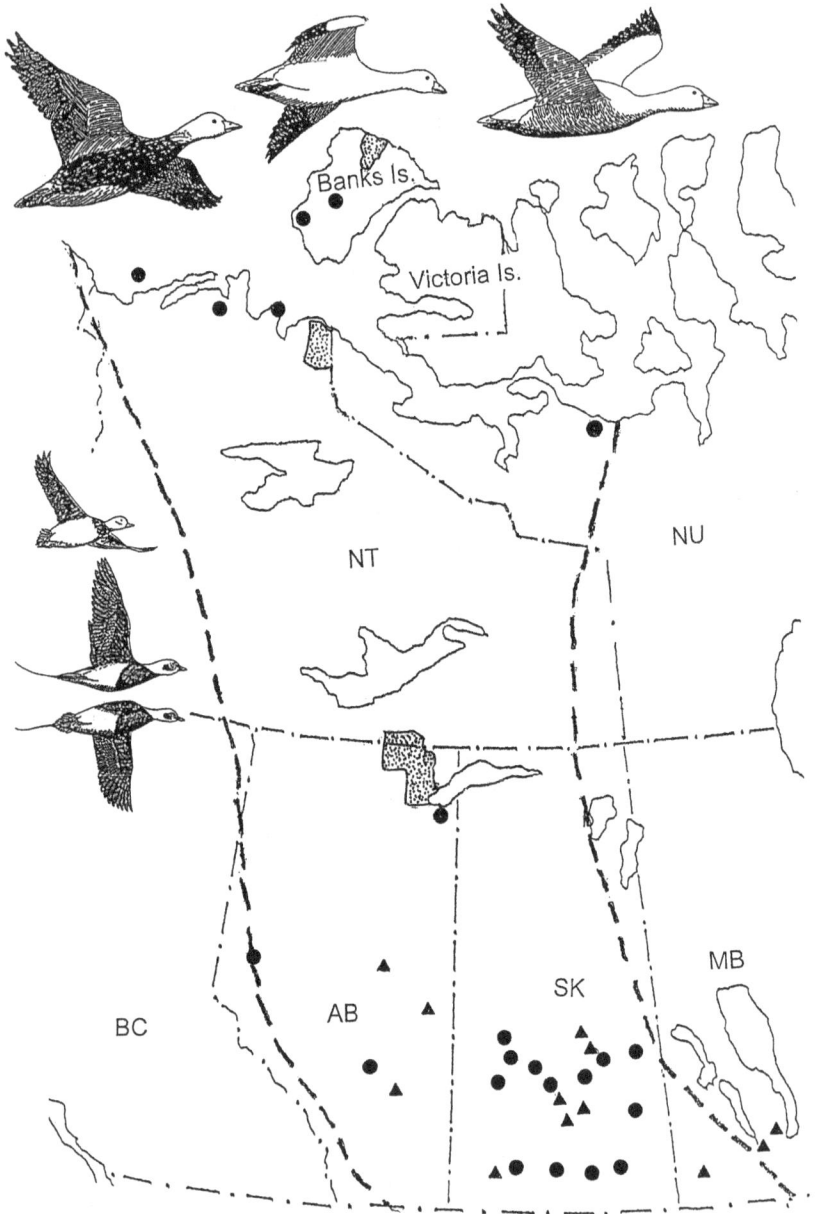

Map 7. Approximate boundaries of the Central Flyway in Canada. National wildlife areas are shown as triangles, migratory bird sanctuaries as circles, and national parks as enclosed and stippled areas. Illustration by author.

where remnants have been preserved in the form of sanctuaries, wild-life management areas and state, provincial or national refuges.

As time has past and natural habitats increasingly degraded or destroyed, preserved sites have become ever more important to migrating birds, a trend that sometimes has resulted in overcrowding, stress, and transmission of diseases. In spite of such limitations, were it not for such sites (Maps 6 &7), our national populations of migrant birds were be in far worse population conditions than is currently the case. This book is in part a testimony to their great present-day importance to migrants (see Chapter 2).

Nearly 100 of the Central Flyway's approximately 400 migrant birds winter in part or entirely within the Neotropic zoogeographic realm, a vast region of the Western Hemisphere that includes all of South and Central America, and those parts of Mexico lying south of the Mexican Plateau. Latin America largely lacks bird sanctuaries, and those that do exist are often wholly lacking in effective protective measures. Preservation of Neotropic migrants and their increasingly ravaged habitats represents one of the greatest perils to the futures of a large proportion of North American migrant birds (Keast & Morton, 1980; Rappole *et al.*, 1983; Buckley *et al.* 1985, Finch & Stangel, 1992; DeGraaf & Rappole. 1995; Martin & Finch, 1995; Stotz *et al.*, 1996).

2

Important Refuges, Sanctuaries, and other Sites of Significance to Migratory Birds in the Central Plains

United States*

COLORADO (Eastern) (CO)

1. Alamosa National Wildlife Refuge. 11,168 acres. Located about seven miles northeast of Alamosa. At least 58 species are common to abundant during spring, *vs.* 48 species during fall, and 16 during winter (Jones, 1990). There are at least 70 nesting birds among the 183 listed for the refuge by Jones. A total of 41 bird species were reported year-around by Jones, so an estimated minimum of 78 percent of the refuge's total bird diversity is migratory. Information is available from the refuge manager, P.O. Box 1148. Alamosa, CO 81101 (303/589-4021). A recent checklist that also includes the birds of Monte Vista N.W.R. is available on-line: http://www.npwrc.usgs.gov/resource/birds/chekbird/r6/8.htm

2. Bonny Lake State Park/South Republican State Wildlife Area. 13,140 acres. Located seven miles south of Idalia. The 1,900-acre Bonny Reservoir on the South Republican River has attracted a great number of migrants in this water-poor region. They include American White Pelicans, Bald Eagles, large flocks of Snow

---

*Adapted in part from Johnsgard (1979, 2012)

Geese and Sandhill Cranes (March and mid-October to mid-November), as well as Tundra Swans, Greater White-fronted Geese, loons, grebes and herons. For information contact the Colorado State Parks Dept. 1313 Sherman St., Room 618, Denver C 89293 (303/866-3437), or the Park (970/354-7306).

3. Comanche National Grassland. 435,707 acres. This region of short-grass prairie is located in the southeaster corner of Colorado, and includes short-grass prairie, mixed-grass prairie, and eroded canyonlands. Little water is present, but about 250 species of birds have been recorded here. Some of the Neotropical migrant species that commonly breed here are the Scissor-tailed Flycatcher, Cassin's Kingbird, Black-chinned Hummingbird, Mississippi Kite, Eastern Phoebe and Dickcissel, the two last-named species at the western edges of their ranges. A bird list is available from the headquarters, 27162 Highway 287, P.O. Box 127, Springfield, CO 81073 (719.523-6591), or from the La Junta office P.O. Box 817, 3rd. St. & E. Hwy. 50, La Junta CO 81050 (719/384-2181).

4. Jackson Lake State Park. 2,540 acres. Located 12 miles north of Wiggins, off I-76. This plains reservoir is an important stopping point of shorebirds, waterfowl, other water birds and Bald Eagles during migration. It also attracts Sandhill Cranes, American White Pelicans, and duck flocks of up to 20,000 birds. Many migrant grassland birds are also present seasonally. For information contact the Colorado State Parks Dept. 1313 Sherman St., Room 618, Denver C 89293 (970-645-2551).

5. John Martin Reservoir State Park. 22,000 acres. Located five miles east of Las Animis, at the confluence of the Arkansas and Purgatoire rivers. This vast area supports shortgrass prairies (with prairie dogs and their associated raptor predators), such as Golden Eagles, Ferruginous Hawks and other buteos. Wintering and migrant waterfowl use the reservoir, as do nesting Great Blue Herons, Double-crested Cormorants and American White Pelicans. The Purgatoire River State Wildlife Area (950 acres) is nearby on the Purgatoire River, with excellent riparian and marshland habitat. For information contact the Colorado Div. of Wildlife, 6060

Broadway, Denver, CO 80216 (303/297-1192), or the Park (719-336-6600).

6. Pawnee National Grassland. 200,000 acres. One entrance to this federally-owned grassland is located two miles north of Briggsdale. Consists of parcels of various sizes, of reclaimed arid grasslands from failed homesteads. Little water is present, and the area is mostly of importance to breeding birds of the short-grass prairie, such as McCown's Longspur and Mountain Plover. A recent checklist of the birds (225 species) of the Pawnee National Grassland is available from the Grassland headquarters, 2000 9th St., Greeley, CO 80631 (303/353-5004), or from the U.S. Forest Service, Bldg. 85, Denver Federal Center, Denver, CO 80225.

7. Prewitt Reservoir State Wildlife Area. 2,900 acres. Located four miles southwest of from Merino, off U.S. Hwy. 6. This is a reservoir along the South Platte River, that attracts Sandhill Cranes, American White Pelicans and many waterfowl during migrations, as well as herons and other water birds, For information contact the Colorado Div. of Wildlife, 6060 Broadway, Denver, CO 80216 (303/297-1192).

8. Queens State Wildlife Area. 4,426 acres. This site is actually made up of several eastern Colorado reservoirs ("Queens reservoirs"): Neenoshe, Neeskah, Neesopah, Negrande, Thurston and Upper Queens. They are scattered near the towns of Lamar and Eades. The reservoirs attract large numbers of Canada, Cackling, Snow and Ross's geese from fall to spring, and a great variety of shorebirds from April through October. For locations and more information contact Colorado Division of Wildlife, 2500 S. Main St., Lamar, CO 81052 (791/336-6600), or on-line at: http://wildlife. state.co.us/landwater/statewildlifeareas

IOWA (Western)(IA)

1. De Soto National Wildlife Refuge. 7,800 acres. Situated on the Nebraska-Iowa border, between Blair, Nebraska, and Missouri Val-

ley, Iowa. The refuge consists of an oxbow cutoff (De Soto Lake) of the Missouri River and encompasses floodplain lands including river-bottom forest and adjacent grasslands, marshes, and cultivated lands. A checklist of 240 species, is available from the refuge manager, R. R. I-B, Missouri Valley, IA 51555 (712/647-4121). It is also available on-line: URL: http://www.npwrc.usgs.gov/resource/birds/chekbird/r3/19.htm

2. Forney Lake Wildlife Management Area. 1,297 acres. This state-owned wetland is located as the base of the loess hills just east of the Missouri river, near the towns of Bartlett and Thurman. It is most noted for its spring concentrations of up to 100,000 Snow Geese and other waterfowl in mid-March, accompanied by up to 100 Bald Eagles. American White Pelicans migrate through about a month later, and various herons and egrets are common through summer. Trumpeter Swans nested here in 2012. Much of the area is used for hunting during fall, and is closed to non-hunters. Managed by Iowa Dept. of Natural Resources (712/374-3133). For information contact the Iowa Dept. of Natural Resources, Wallace State Office Bldg., Des Moines, IA 503129-0034 (515/282-5145).

4. Hitchcock Nature Area. 582 acres. This county-operated area, located about eight miles north of Crescent, lies at the base of the loess hills and east of the Missouri River. The loess hills, up to about 200 feet high, provide a natural migration route for raptors that use the updraft winds for lift, and form a narrow migration corridor used by thousands of hawks, vultures and other raptors annually. During one fall over 6,900 raptors of 19 species were counted, of which the eight most common species were Red-tailed Hawk (41%), Swainson's Hawk (22%), Sharp-shinned Hawk (14%), Bald Eagle (5%), Broad-winged Hawk (4%), Northern Harrier (3%), Cooper's Hawk (3%) and American Kestrel (2.5%). Large numbers of Turkey Vultures also pass through. For information contact the Pottawattamie County Conservation Board (712/328-5638).

3. Riverton Wildlife Management Area. 2,700 acres. This is a large,

state-owned wetland located about a mile north of Riverton in the Nishnabotna River Valley, and is notable for its large concentrations of migratory waterfowl, especially Snow Geese, as well as their frequently associated Bald Eagles. Waterfowl numbers peak in March and again in late November to mid-December, which Snow Goose numbers may reach 200,000 and the eagles up to 50 or more. Depending on spring water levels, up to 20 or more species of shorebirds may be present including Hudsonian Godwit, Baird's and Stilt sandpipers, and Wilson's Phalarope. Managed by Iowa Dept. of Natural Resources (712/374-32133). For information contact the Iowa Dept. of Natural Resources, Wallace State Office Bldg., Des Moines, IA 503129-0034 (515/282-5145).

4. Union Slough National Wildlife Refuge. 2,845 acres. This refuge consists of marsh, grassland and timber. At least 91 species are common to abundant during spring, *vs.* 97 species during fall, and 18 during winter (Jones, 1990). During March and April thousands of waterfowl pass through, especially Canada and Snow geese, plus Mallards, Blue-winged and Green-winged teal, Gadwall and American Wigeon. Other common migrants are American White Pelicans, Bald Eagles, Franklin's Gulls, Forster's Terns, and Great Egrets, as well as many shorebirds and passerines. There are at least 96 nesting birds among the 217 species listed for the refuge by Jones. A total of 30 bird species were reported present year-around by Jones, so an estimated minimum of 86 percent of the refuge's total bird diversity is migratory. A recent checklist of more than 240 species, is available from the refuge manager, Rte. 1., Box 52, Titonka, IA 50480 (515/928-2523). It is also available on-line: URL: http://www.npwrc.usgs.gov/resource/birds/chekbird/r3/19.htm

KANSAS (KS)

1. Cheyenne Bottoms Waterfowl Management Area. 18,000 acres. This famous state-owned wildlife area is about five miles north of Great Bend, in Barton County. It consists of marshland as well as adjacent bottomlands associated with the Arkansas River.

The site is recognized as being of international importance for migratory shorebirds, at peak holding as many as 200,000 or more migrants. Notable breeding species include the Least Bittern, Yellow-crowned Night-heron, King Rail, Common Gallinule and Snowy Plover. The site's birding attractions were described by Zimmerman and Patti (1988), and also by Gress and Janzen (2008). For general information, contact Kansas Dept. of Wildlife & Parks (316/793-7730). A checklist of 325 species was included by Zimmerman (1990) in his book on the area's ecology. Among the long-distance migrants are 34 species of waterfowl, 39 shorebirds, 12 flycatchers, six vireos and 24 warblers. A bird list is available from the area manager, Route 1, Great Bend, KS 67530. There is no on-line bird checklist, but the site's URL is: http://www.cheyennebottoms.net

2. Cimarron National Grassland. 108,175 acres. This multi-parcel national grassland (mixed short-grass prairie and sagebrush–yucca prairie) is located a few miles north of Elkhart, at the southwestern corner of Kansas. Considering the region's aridity, a remarkably long bird list of 342 species was published by Cable, Seltman and Cook (1996). Among them are 27 waterfowl, 35 shorebirds, 9 gulls and terns, 9 herons and egrets, and 16 hawks, eagles and falcons. Abundant overwintering non-breeding species include the Lapland Longspur, American Tree and White-crowned sparrows, Dark-eyed Junco and Pine Siskin. Common breeding short-distance migrants or permanent residents include the Ring-necked Pheasant, Mourning Dove, Horned Lark, Cassin's Sparrow, Grasshopper Sparrow, Red-winged Blackbird and Western Meadowlark. Breeding Neotropical migrants include the Mississippi Kite, Swainson's Hawk, Common Nighthawk, Chimney Swift and Yellow-billed Cuckoo, plus three swallows, seven vireos, 35 warblers, three tanagers and two orioles. Three breeding bird survey routes revealed 58 species. Headquarters address: 737 Villymaca. Elkhart, KS 67950, or on-line: http://www. fs.fed.us/r2/psicc/cim (316/697-4621).

3. Flint Hills National Wildlife Refuge. 18,500 acres. This refuge is on the upper end of the John Redmond Reservoir of the Neo-

sho River in Coffey County. Most of the refuge consists of the reservoir itself, and is managed primarily for waterfowl. Notable breeding species include Wood Duck, Least Bittern and Upland Sandpiper. During migration up to 100,000 waterfowl may be present. Many waterfowl overwinter here, which attract substantial numbers of Bald Eagles. At least 123 species are common to abundant during spring, *vs.* 114 species during fall, and 38 during winter (Jones, 1990). There are at least 86 nesting birds among the 285 listed for the refuge by Jones. A total of 62 bird species were reported present year-around by Jones, so an estimated minimum of 78 percent of the refuge's total bird diversity is migratory. The site's birding aspects were described by Zimmerman and Patti (1988), Gress and Potts (1993), and by Gress and Janzen (2008). A recent checklist of more than 290 species, including 113 wetland species (21 breeders), is available from the refuge manager, P.O. Box 128, Hartford, KS 66854 (Ph. 316/392-5553). It s also available on-line: http://www.npwrc.usgs.gov/resource/birds/chekbird/r6/20.htm

4. Gardner Wetlands (Kansas City Power & Light Company Wetland Park). 55 acres. This suburban site was described by Gress and Janzen (2008). It is notable for its migratory shorebirds and other migratory species that use a 23-acre wetland. Located just west of Gardner, on South Waverly Rd., in Johnson County.

5. Kirwin National Wildlife Refuge. 10,800 acres. This refuge is about ten miles southeast of Phillipsburg, in Phillips County. It consists mostly of marshes, grasslands, croplands and a 5,000-acre reservoir impounded by the north fork of the Solomon River. Large numbers of ducks (especially Mallards) and Canada Geese winter here. Flocks of migrating Sandhill Cranes regularly stop here, and rarely Whooping Cranes are seen. At least 60 species are common to abundant during spring, *vs.* 61 species during fall, and 20 during winter (Jones, 1990). There are at least 46 nesting birds among the 191 listed for the refuge by Jones. A total of 56 bird species were reported present year-around by Jones, so an estimated minimum of 71 percent of the refuge's total bird diversity is migratory The site's birds and birding opportunities

were described by Zimmerman and Patti (1988), and by Gress and Janzen (2008). A recent checklist of 234 total species, including 131 wetland species (10 breeders, mostly ducks and the Least Tern), from the refuge manager, Rte 1, Box 103, Kirwin, KS 67644 (913/543-6673). It is also available on-line: http://www.npwrc. usgs.gov/resource/birds/chekbird/r6/20.htm

6. Marais des Cygnes National Wildlife Refuge and Marais des Cygnes Wildlife Area. 15,000 acres collectively. These two adjoining areas encompass a wide diversity of wetlands ("marsh of the swans"), prairie, deciduous upland and riparian woodland, and transitional habitats. Reflecting this diversity, at least 321 bird species, including 117 breeders, have been identified here, but no published bird list for these sites is yet available. More than 30 warblers have been seen here; breeding species include Yellow-throated, Kentucky, Black-and-White, American Redstart, Parula, Prothonotary, Yellow-breasted Chat and Cerulean (rare). These two sites and their birds were described by Gress and Janzen (2008). Address: Rt. 2, Box 185A, Pleasanton, KS 68075. The refuge is administered by Flint Hills N.W.R. (316/392-5553). For the state-owned wildlife area, phone 913/351-8941.

7. McPherson Valley wetlands. 1,310 acres. This state-owned area is located northeast of Inman, in McPherson County. It includes Lake Inman, the largest natural lake in Kansas, as well as relict marshes formed at the end of the Pleistocene. The site is important for migrating water birds and was described by Gress and Potts (1993). For more information contact Kansas Dept. Wildlife & Parks (316-767-5900). No bird list is yet available.

8. Neosho Wildlife Area. 3,246 acres. This state-owned wildlife area is the largest wetland in southeastern Kansas, and consists of wetlands and riparian woodlands in the Neosho River valley and is formed by putting levees in an old oxbow. It is especially important for migrating waterfowl and shorebirds, and has been described by Gress and Potts (1993), and by Gress and Janzen (2005). Summer resident wading birds include Snowy, Great and Cattle egrets, as well as both night-herons. Notable breed-

ing Neotropical passerines include Acadian Flycatcher, Yellow-throated Vireo. Prothonotary Warbler, Northern Parula and Summer Tanager. Located one mile east of St. Paul, in Neosho County. For more information contact Kansas Dept. Wildlife & Parks, Headquarters Office, Rte. 2, Box 51A, Pratt, KS 67124 (326/672-5911, or 316/362-3671), or on-line at: http://www.kdwp. state.ks.us

9. Perry Reservoir and Perry Lake State Park. This site in the Delaware River Valley consists of Perry Reservoir (11,000 acres), marshes, mudflats, prairie, old fields and riparian woodlands. The area and its birds have been described by Gress and Potts (1993) and by Gress and Janzen (2005). Its associated wetlands of importance to birds include Kyle, Lassiter and Ferguson marshes. Summering migrants include a variety of buntings, vireos, tanagers, thrushes, and warblers, while during fall and spring migrating ducks, Snow Geese and Bald Eagles are abundant. Address of reservoir: U.S. Army Corps of Engineers, 10419 Perry Park Drive, Perry, KA. 66073 (785/597-5144). Address of Perry State Park, 5441 Westlake Rd., Ozawkie, KS 66070 (913/246-3449). No bird list is yet available.

10. Quivira National Wildlife Refuge. 21,800 acres. This outstanding marshland refuge is located 12 miles northeast of Stafford, Stafford County. It consists of i4,700 acres of marsh, as well as grassland, farmlands, and sandhills. It is notable for its diverse wetland birds, including consistent use by both Sandhill and Whooping cranes. Notable breeding species include Hooded Merganser, Eared Grebe, Least Bittern, Great Egret, Snowy Egret, Yellow-crowned Night-heron, White-faced Ibis, Black, King, Sora and Virginia rails, Snowy Plover and Black-necked Stilt. At least 112 species are common to abundant during spring, *vs.* 110 species during fall and 39 during winter (Jones, 1990). There are at least 88 nesting birds among the 252 listed for the refuge by Jones. A total of 67 bird species were reported present year-around by Jones, so an estimated minimum of 74 percent of the refuge's total bird diversity is migratory. A recent checklist of 340 species is available from the refuge manager, R.R. 3, Box 48A,

Stafford, KS 67578 (Ph. 316/486-2393). It is also available on-line: http://www.npwrc.usgs.gov/resource/birds/chekbird/r6/20.htm

11. Tuttle Creek Lake area. 28,500 acres. This largest of Kansas reservoirs is north of Manhattan, and has public lands and wetlands associated with Tuttle Creek Lake. Significant wetlands with public access include Fancy Creek Wildlife Area, Olsburg Marsh (north of Olsburg on Shannon Creek Road), Carnahan Creek Park, Outlet Park, Tuttle Creek Park, Stockdale Park, and River Pond State Park. The last-named area is noted for its winter Bald Eagle population. Information on these sites can be obtained from the Corps of Engineers Visitor Center at Tuttle Creek Dam (785/539-8511). No bird list is yet available.

MINNESOTA (Western) (MN)

1. Agassiz National Wildlife Refuge. 61,500 acres. Situated 11 miles east of Holt, 490 in Marshall County, Minnesota. This area, once a part of glacial Lake Agassiz, contains grasslands with hardwood groves, potholes, and lakes, at the eastern transition zone between tallgrass prairie and northern forest. At least 86 species are common to abundant during spring, vs. 69 species during fall, and 8 during winter (Jones, 1990. There are at least 132 nesting birds among the 248 listed for the refuge by Jones. A total of 18 bird species were reported present year-around by Jones, so an estimated minimum of 93 percent of the refuge's total bird diversity is migratory. A recent checklist of 287 species reported on the refuge is available from the refuge manager, Middle River, MN 56737 (218/449-4115). It is also available on-line: http://www.npwrc.usgs.gov/resource/birds/chekbird/r3/27.htm

2. Big Stone National Wildlife Refuge. 10,000 acres. Located on the Minnesota River near the South Dakota border, with over 1,700 aces of native prairie. Information is available from the refuge manager, 25 NW 2nd. St., Ortonville, MN 56278. At least 101 species are common to abundant during spring, vs. 89 species during fall, and 15 during winter (Jones, 1990). The area is an impor-

tant stopover for migrant warblers in spring. There are at least 107 nesting birds among the 237 listed for the refuge by Jones. A total of 57 bird species were reported present year-around by Jones, so an estimated minimum of 76 percent of the refuge's total bird diversity is migratory. A recent checklist of 240 species reported on the refuge is available from the refuge manager, 25 Northwest 2nd. St., Ortonville, MN 56278 (512/839-3700). It is also available on-line: http://www.npwrc.usgs.gov/resource/birds/chekbird/r3/27.htm

3. Fergus Falls Wetland Management District. Located at the edge of the Red River valley, and the shoreline area of glacial Lake Agassiz. A recent checklist of 290 species reported on this multi-county W.M.D. is available from the district manager, Rte. 1, Box 76, Fergus Falls, MN 56537, It is also available on-line: http://www.npwrc.usgs.gov/resource/birds/chekbird/r3/27.htm

4. Heron Lake Wetlands. This famous wetland, located between the towns of Heron Lake and Lakeland, is one of the largest prairie wetlands in North America. In addition to attracting great numbers of migrating waterfowl, it is a breeding area for Western Grebes, Forster's and Black terns, Canada Geese, American White Pelican, and Double-crested Cormorants. Franklin's Gulls once had an enormous nesting colony here, until fluctuating water levels destroyed their vegetation base. For information contact DNR Wildlife Area Office, Windom, MN 56101 (507/831-2917).

5. Minnesota Waterfowl Production Areas. A recent checklist of 266 species reported from waterfowl production areas in the prairie wetlands around Fergus Falls, northwestern Minnesota, is available from the manager, Minnesota Wetland Complex Office, Rte. 1, Box 76. Fergus Falls, MN 56537. It is also available on-line: http://www.npwrc.usgs.gov/resource/birds/chekbird/r3/27.htm

6. Lac Qui Parle Wildlife Management Area/State Park. 31,238 acres. Located between Appleton and Montevideo, on the Minnesota River. It is a major stopping point for Canada Geese (up to

150,000), Tundra Swans. Snow Geese, and other migrating waterfowl, and a nesting area for American White Pelicans (up
to 10,000), Bald Eagles and many prairie birds. Adjacent to the
WMA is Plover Prairie, with nesting Upland Sandpipers and
Marbled Godwits. Nearby on the South Dakota border south of
Marietta is Salt Lake Wildlife Management Area. This is the largest saline wetland in Minnesota, and a magnet for shorebirds
such as American Avocets, which sometimes nest. Over 140 bird
species have been identified here. For information contact DNR
Wildlife Area Office, Lac Qui Parle WMA, RR 1 Box 23, Watson,
MN 56295 (320/734-45451).

7. Morris Wetland District. A recent checklist 282 species reported in
   this multi-county wetland district is available from the district
   manager, 43875 230th. St., Morris, MN 56267. It is also available
   on-line: http://www.npwrc.usgs.gov/resource/birds/chekbird/
   r3/27.htm

8. Roseau River Wildlife Management Area. 62,025 acres. Located
   20 miles northwest of Roseau. This is one of Minnesota's most
   important waterfowl migration and breeding areas, with 149
   known breeding species, and an important stopover area for arc-
   tic-bound Sandhill Cranes, Tundra Swans and shorebirds. For information contact DNR Wildlife Area Office, HCT 5 Box 103, Roseau MN 56751 (218/463-1557).

9. Talcot Lake Wildlife Management Area. 4,000 acres. Located between Westbrock and Dundee. This lake and marshy wetland
   on the Des Moines River attracts large flocks of waterfowl, especially Canada Geese of the eastern prairie flock, during fall migration. The area is one of the largest protected wildlife areas in
   southwestern Minnesota, and a magnet for marsh and prairie
   birds. For information contact DNR Wildlife Area Office, Talcot
   Lake WMA. RR 3 Box 534, Dundee, MN 56131 (507/468-2248).

MISSOURI (Western) (MO)

1. Squaw Creek National Wildlife Refuge. 6,887 acres, This refuge is five miles south of Mound City, Holt County, in extreme north-western Missouri. It consists of marshes, Missouri River bot-tomlands, wooded bluffs, and farmlands. At least 90 species are common to abundant during spring, *vs.* 71 species during fall, and 25 during winter (Jones, 1990). There are at least 104 nest-ing birds among the 268 listed for the refuge by Jones. A total of 65 bird species were reported present year-around by Jones, so an estimated minimum of 76 percent of the refuge's total bird di-versity is migratory. Depending on the severity of the winter, up to 400,000 Snow Geese, other geese, ducks, and up to 100 or more Trumpeter Swans may overwinter here. A recent checklist of 268 species is available from the refuge manager, Box 101, Mound City, MO 64470 (816/442-3187). It is also available on-line: http://www.npwrc.usgs.gov/resource/birds/chekbird/r3/29.htm

2. Swan Lake National Wildlife Refuge. 10, 479 acres, Located just east of Sumner. At least 103 species are common to abundant during spring, *vs.* 86 species during fall, and 28 during winter (Jones, 1990). There are at least 85 nesting birds among the 233 listed for the refuge by Jones. A total of 48 bird species were re-ported present year-around by Jones, so an estimated minimum of 80 percent of the refuge's total bird diversity is migratory. A recent checklist of 237 species is available from the refuge man-ager, P.O. Box 68, Sumner, MO 65681 (816/856-3323). It is also available on-line: http://www.npwrc.usgs.gov/resource/birds/chekbird/r3/29.htm

MONTANA (Eastern)(MN)

1. Benton Lake National Wildlife Refuge. 10.383 acres. Located 14 miles north of Great Falls. At least 33 species are common to abundant during spring, *vs.* 35 species during fall (Jones, 1990). There are at least 59 nesting birds among the 167 listed for the refuge by Jones. A total of 10 bird species were reported present

year-around by Jones, so an estimated minimum of 94 percent of the refuge's total bird diversity is migratory. Some of the major migrants are Snow Goose, Tundra Swan, Northern Pintail, American Wigeon, and many shorebirds. The refuge has been recognized as an internationally significant shorebird site by the Western Hemisphere Shorebird Reserve Network. Nesting shorebirds include Upland Sandpiper, Marbled Godwit and Willet. A recent checklist of 199 species is available from the refuge manager, 922 Bootlegger Rd., Great Falls, MT 59404 (406/727-7400). It is also available on-line: http://www.npwrc.usgs.gov/resource/birds/chekbird/r6/30.htm

2. Bowdoin National Wildlife Refuge. 15,337 acres. Located seven miles east of Malta. At least 98 species are common to abundant during spring, *vs.* 87 species during fall, and 10 during winter (Jones, 1990). There are at least 102 nesting birds among the 266 listed for the refuge by Jones. A total of 9 bird species were reported present year-around by Jones, so an estimated minimum of 97 percent of the refuge's total bird diversity is migratory. A recent checklist of 248 species is available from the refuge manager, PO Box J, Malta, MT 59538 (406/654-2863). It is also available on-line: http://www.npwrc.usgs.gov/resource/birds/chekbird/r6/30.htm

3. Charles M. Russell National Wildlife Refuge & UL Bend National Wildlife Refuge. 1,094,301 acres. At least 15 species are common to abundant during spring, *vs.* 101 species during fall, and 20 during winter (Jones, 1990). There are at least 98 nesting birds among the 252 listed for the refuge by Jones. A total of 35 bird species were reported present year-around by Jones, so an estimated minimum of 86 percent of the refuge's total bird diversity is migratory. A recent checklist of 240 species is available from the refuge manager, P.O. Box 110, Lewistown, MT 59457 (4-6/538-8707). It is also available on-line: http://www.npwrc.usgs.gov/resource/birds/chekbird/r6/30.htm

4. Freezout Lake Wildlife Management Area. 11,350 acres. Located ten miles southeast of Choteau. This area is mainly of in-

terest for its great spring concentrations of migrating waterfowl, which peak in late March, when up to 300,000 Snow and
Ross's geese from wintering grounds in California and the southern Great Plains stage prior to leaving for breeding grounds in
the northwestern corner of Northwest Territories There are also
up to 12,000 Tundra Swans, and many migrating ducks, at times
totaling a million waterfowl. The area is hunted during fall, so
the numbers are less impressive then, but Tundra Swans move
through in good numbers during late October and early November. For information contact the Montana Dept. of Fish, Wildlife
& Parks, PO Box 488, Fairfield, MT 58426 (406/467-2646).

5. Medicine Lake National Wildlife Refuge. 31,457 acres. Located one
   mile south of Medicine Lake. At least 69 species are common to
   abundant during spring, vs. 56 species during fall, and 6 during
   winter (Jones, 1990). There are at least 96 nesting birds among
   the 219 listed for the refuge by Jones. A total of 15 bird species
   were reported present year-around by Jones, so an estimated
   minimum of 93 percent of the refuge's total bird diversity is migratory. Migrants that regularly visit the refuge are Sandhill and
   Whooping cranes, Tundra Swan, and many other waterfowl and
   shorebirds. The areas around the lakes are a mixture of tallgrass
   and short-grass prairies, with associated birds, A recent checklist of 228 species is available from the refuge manager, HC 51,
   Box 2, Medicine Lake, MT 59247 (406/799-2305). It is also available on-line: http://www.npwrc.usgs.gov/resource/birds/chek-
   bird/r6/30.htm

NEBRASKA (NE)

1. Central Platte Valley. This approximately 80-mile stretch (the "Big
   Bend") of the Platte River in central Nebraska, roughly between
   Lexington and Grand Island, hosts one of the world's great bird
   spectacles in early spring, when millions of waterfowl and a half-
   million Sandhill Cranes descend into the Platte Valley and the
   adjoining Rainwater Basin immediately to the south. Estimates of
   waterfowl variety greatly, and actual numbers depend on water

conditions, but it is commonly estimated that up to nine million
birds, including up to an estimated maximum of seven million
Snow Geese (but more often 1-2 million), may be here in mid-
March, along with hundreds of thousands each of Canada Geese
and Greater White-fronted Geese, probably 50-100,000 Cackling
Geese, and perhaps 5-10,000 Ross's Geese, the latter two species'
population sizes still poorly documented. The numbers of ducks
are equally impressive, with Mallards and Northern Pintails perhaps the most common and earliest to arrive, competing with
the cranes and geese for unharvested corn. By the end of March,
20 or more duck species will have arrived, while the geese and
Sandhill Cranes will have begun to leave. By mid-April the
Sandhill Cranes (mostly arctic tundra-breeding lesser Sandhills
headed for Alaska and Siberia), are being replaced by the earliest of the shorebird migrants and small groups of Whooping
Cranes. The shorebirds usually peak by the end or April or mid-
May, their totals being estimated at 200,000–300,00 birds. The
birds of this 10,000 square-mile region, which total over 390 species, have been documented by Brown and Johnsgard (in prep).
There are no federal refuges on this stretch of the Platte River,
but one important sanctuary is the 1,600-acre Rowe Sanctuary
and Ian Nicolson Audubon Center, about three miles south of
Gibbon. This sanctuary protects over seven miles of prime crane
feeding and roosting habitat, and up to 70,000 roosting cranes
can often be seen from its blinds. Least Terns and Piping Plovers often nest on barren sandbars that are also used by roosting
cranes. Summer breeding birds include Dickcissel, Upland Sandpiper and Bobolink, as well as riparian wooded habitats species
such as Rose-breasted Grosbeak and Willow Flycatcher. Rowe is
also immediately north of the western Rainwater Basin wetlands
(see below) For information contact Rowe Sanctuary, 44450 Elm
Island Road, Gibbon, NE 68840 (308/468-5282). The Crane Trust
(308/384-4633), a mile east of the Alda Road bridge on Whooping Crane Drive, controls and manages several thousand acres of
riparian wetlands along the Platte River. It's biologists have performed annual surveys of Sandhill Crane usage, habitat surveys,
and other avian and ecological research. Through its associated
Crane Trust Visitor and Nature Center (308/382-1820), the Trust

also provides information to tourists and, like Rowe Sanctuary, offers visitors access to crane-viewing blinds in the vicinity of Alda Bridge. See websites at http://CraneTrust.org_ and http://NebraskaNature.org_

2. Crescent Lake National Wildlife Refuge. 45,818 acres. Located 28 miles north of Oshkosh, via gravel and unimproved roads. There are about 20 wetland complexes on this enormous Sandhills refuge; the wetlands total 8,251 acres, and comprise almost 20 percent of the refuge. At least 32 species of waterfowl have been reported here, and 14 are known or suspected breeders. Three grebes (Western, Eared and Pied-billed) are also breeders. Other wetland breeders include the Double-crested Cormorant, Great Blue Heron, Black-crowned Night-heron, Sora and Virginia Rail, and Black and Forster's terns. The Common Yellowthroat and Sedge and Marsh Wrens are abundant, and both the White-faced Ibis and Black-necked Stilt now breed regularly. The marshes and shallow lakes in this large and remote Sandhills refuge vary greatly as to their relative alkalinity. Border Lake at the western edge of the refuge marks the eastern boundary of hypersaline water conditions; the Wilson's Phalarope and American Avocet are common breeders here. The refuge and its birds have been described by Farrar (2004) and Johnsgard (1995, 2011). At least 66 species are common to abundant during spring, *vs.* 56 species during fall and 6 during winter (Jones, 1990). There are at least 83 nesting birds among the 233 listed for the refuge by Jones. A total of 40 bird species were reported present year-around by Jones, so an estimated minimum of 83 percent of the refuge's total bird diversity is migratory. The refuge bird list includes 273 species, with many wetland species, and is available from the refuge manager, 10630 Rd. 181, Ellsworth, NE 69340 (Ph. 308-762-4893). It is also available on-line: http://www.npwrc.usgs.gov/resource/birds/chekbird/r6/31.htm The refuge URL is: http://crescentlake.fws.gov/

3. De Soto National Wildlife Refuge. See Iowa listing for refuge description,

4. Fort Niobrara National Wildlife Refuge. 19,122 acres. Includes about 4,350 acres of mostly riparian woods and 375 acres of wetlands. Riparian hardwood forest along the Niobrara River and upland Sandhills prairie, with some spring-fed ponds. Notable wetland species include the Wood Duck, Upland Sandpiper, and Long-billed Curlew. The most abundant Neotropical migrants nesting in the refuge area are the Common Yellowthroat, Red-eyed Vireo, Ovenbird and Black-and-White Warbler. Located about five miles east of Valentine along State Highway 12 (Ph. 402/376-378). At least 71 species are common to abundant during spring, *vs.* 67 species during fall and 14 during winter (Jones, 1990). There are at least 76 nesting birds among the 201 listed for the refuge by Jones. A total of 25 bird species were reported present year-around by Jones, so an estimated minimum of 88 percent of the refuge's total bird diversity is migratory. The most recent refuge bird list includes 230 species, many of which are riparian woodland species with primarily eastern zoogeographic affinities. It is available from the refuge manager, Hidden Timber Rte., HC 14, Box 67, Valentine, NE 69201 (402/376-3789). An on-line refuge list is also available: http://www.npwrc.usgs.gov/resource/birds/chekbird/r6/31.htm The refuge website is: http://fortniobrara.fws.gov/

5. Lake McConaughy State Recreation Area. 41,192 acres. This largest of Nebraska's reservoirs (over 30,000 acres), was developed for flood control, irrigation and recreational use. It is about 22 miles long, 3 miles wide, up to 140 feet deep, and has 105 miles of shoreline when full. Including the adjacent Lake Ogallala State Recreation Area, the site totals about 5,500 land acres. Mostly bare sandy shorelines are on northern side, but extensive wetlands exist at Clear Creek Wildlife Management Area, at the lake's western end. The lake is an important nesting area for both Piping Plovers and Least Terns. The lake also hosts many non-breeding Double-crested Cormorants and American White Pelicans throughout summer. Western and Clark's grebes summer and breed here; up to about 20,000-30,000 Western Grebes stage here during fall migration. Large numbers of waterfowl, gulls, other water birds and eagles winter here. Located nine miles

north of Ogallala on State Highway 61. State park entry permit required (Ph. 308/284-8800). A recent checklist for the Lake Mc-Conaughy region has 362 species (Brown, Dinsmore and Brown, 2012), a high percentage of which are wetland-dependent species. The area's website is: http://www.lakemcconaughy.com/ngp.html

6. North Platte National Wildlife Refuge. 5,047 acres. Part of the Crescent Lake/North Platte N.W.R. complex, and including Lake Alice (1,377 acres when full, but recently dry), Lake Minatare (430 acres) and Winters Creek (700 acres). The best wetland bird habitat is at Winters Creek, which seasonally supports many waterfowl and Sandhill Cranes. Located four miles north and eight miles east of Scottsbluff, Scotts Bluff County. The refuge bird list totals 228 species, including 85 wetland species (13 breeders), and is available on-line or from the refuge manager, 10630 Road 181, Ellsworth, NE 69340 (308-762-4893). It is also available on-line: http://www.npwrc.usgs.gov/resource/birds/chekbird/r6/31.htm

7. Oglala National Grassland 94,344 ac. Shortgrass prairie and eroded badlands/ Bird list of 302 pp. (covering all of Pine Ridge area) available from U.S. Forest Service 270 Pine St., Chadron, NE 69337. Ph. 308/432-3367 or 308/432-4475.

8. Rainwater Basin Wetland Management District. This multi-county region south of the central Platte River (see above) contains hundreds of temporary to seasonal playa wetlands. The Rainwater Basin Joint Venture coordinates the Rainwater Basin's wetland management, which involve the approximately 50 federally owned waterfowl production areas, and about 30 state-owned wildlife management areas, extending from Phelps County east to Butler and Saline counties, geographically divided into eastern and western components. The Rainwater Basin's importance to Great Plains migrating shorebirds during April and early May is probably second only in the Great Plains to Cheyenne Bottoms in Kansas, and during wet springs it also often holds millions of migrating geese (mostly Snow Geese, Greater White-fronted Geese and Canada Geese) during March. Nebraska's playa wet-

lands are included within the multi-state Playa Lakes Joint Venture program, which extends from northwestern Nebraska south to western Texas. At least 29 species are common to abundant during spring, *vs.* 122 species during fall, and 28 during winter (Jones, 1990). There are at least 102 nesting birds among the 256 listed for the district by Jones. A total of 49 bird species were reported present year-around by Jones, so an estimated minimum of 81 percent of the district's total bird diversity is migratory. The region and its birds have thoroughly been described by Jorgensen (2012), whose report is available on the Nebraska Game & Park Commission's website (see references). The address for the Rainwater Basin Wetland Management District is: 73746, P. O. Box 8, Funk, NE 68940 (308/263-3000).The Rainwater Basin Joint Venture's address is 2550 N. Diers Ave., Suite L, Grand Island, NE 68803 (308/382-8112). The Playa Lakes Joint Venture's address is 103 East Simpson St., LaFayette, CO 80026 (Ph. 303-926-0777). A collective bird list for the Rainwater Basin and adjacent central Platte Valley has more than 300 species, including 120 wetland species, and is available from the U.S. Fish & Wildlife Service, P.O. Box 1766, Kearney, NE 68847. It is also available on-line: http://www.npwrc.usgs.gov/resource/birds/chekbird/r6/plandrwb.htm

9. Valentine National Wildlife Refuge. Area 71,516 acres. Located 22 miles south of Valentine. Nebraska's largest national wildlife refuge, consisting mostly of Sandhills prairie, with sand dunes and intervening depressions that contain many shallow, sometimes lake-sized, marshes. It covers more than 71,000 acres and includes 36 lakes plus numerous marshes, surrounded by sand dunes up to 200 feet high. Many migrant grassland species, such as Long-billed Curlew and Upland Sandpiper, are abundant on this enormous refuge. Two curlews with radio transmitters banded at Valentine N.W.R. were found to winter on the Gulf Coast of northeastern Mexico, returning to their prior nesting area the following spring. Four grebes (Eared, Western, Clark's and Pied-billed) nest here, as do the White-faced Ibis, Long-billed Curlew, Upland Sandpiper, Wilson's Phalarope and American Avocet. Up to 150,000 migrant ducks can be found on the ref-

uge, with peak numbers occurring in May and October. At least 67 species are common to abundant during spring, *vs.* 66 species during fall and 8 during winter (Jones, 1990). There are at least 95 nesting birds among the 233 listed for the refuge by Jones. A total of 35 bird species were reported present year-around by Jones, so an estimated minimum of 85 percent of the refuge's total bird diversity is migratory. The most recent refuge checklist of 272 species includes 100 wetland species, including 31 shorebirds, 24 waterfowl, 10 gulls and terns, 5 grebes and 4 rails. It is available from the refuge manager, Hidden Timber Rte., HC 14, Box 67, Valentine, NE 69201. (402/376-378). It is also available on-line: http://www.npwrc.usgs.gov/resource/birds/chekbird/r6/31.htm The refuge URL is http://valentine.fws.gov/

NEW MEXICO (Eastern) (NM)

1. Bitter Lake National Wildlife Refuge. 23,310 acres. Located ten miles northeast of Roswell. At least 68 species are common to abundant during spring, *vs.* 60 species during fall, and 35 during winter (Jones, 1990). A recent checklist of 285 species recorded on the refuge is available from the refuge manager, P.O, Box 7, Roswell. NM 88201 (505/611-6755). It is also available on-line: http://www.npwrc.usgs.gov/resource/birds/chekbird/r2/35.htm

2. Bosque del Apache Wildlife Refuge. 57,191 acres. Located 93 miles south of Albuquerque. At least 83 species are common to abundant during spring, *vs.* 76 species during fall, and 55 during winter (Jones, 1990). There are at least 95 nesting birds among the 252 listed for the refuge by Jones. A total of 83 bird species were reported present year-around by Jones, so an estimated minimum of 67 percent of the refuge's total bird diversity is migratory. This is a major wintering area for the Rocky Mountain population of greater Sandhill Cranes, as well as for Snow and Ross's geese. A checklist of 377 species recorded on the refuge is available from the refuge manager, P.O, Box 1246, Socorro, NM78701 (505/835-1828). It is also available on-line: http://www.npwrc.usgs.gov/resource/birds/chekbird/r2/35.htm

3. Grulla National Wildlife Refuge. 3,235 acres. This refuge is located a few miles east of Arch. Most of it is covered by a salt lake, with the rest grassland. It largely managed to provide winter habitat for Sandhill Cranes, and 100,000 or more may be present by December. Shorebird numbers and diversity depend on water conditions. At least 46 species are usually common to abundant during spring, vs. 60 species during fall and 31 during winter (Jones, 1990). There are at least 21 nesting birds among the 88 listed for the refuge by Jones. A total of 42 bird species were reported present year-around by Jones, so an estimated minimum of 52 percent of the refuge's total bird diversity is migratory. A checklist of 85 species recorded on the refuge is available from the refuge manager, c/o Muleshoe N.W.R., P.O. Box 549, Muleshoe, TX 79347 (806-946-3341). It is also available on-line: http://www. npwrc.usgs.gov/resource/birds/chekbird/r2/35.htm

4. Kiowa National Grassland 136,505 ac. Shortgrass upland plains. Bird list available (226 spp., 60 residents or summer residents). Address: 16 N. 2nd St., Clayton. NM 88415. Ph. 505/374-9652.

5. Rita Blanca National Grassland. See Texas listing.

NORTH DAKOTA (ND)

1. Arrowwood National Wildlife Refuge. 15,934 acres. Situated about 14 miles north of Jamestown, North Dakota. Contains lakes, marshes, grasslands, wooded areas and fields. At least 64 species are common to abundant during spring, vs. 59 species during fall and 25 during winter (Jones, 1990). There are at least 105 nesting birds among the 246 listed for the refuge by Jones. Among them are five species of grebes, 13 ducks, Forster's and Black terns, Willets, Marbled Godwits, American Avocets and Wilson's Phalaropes. A total of 19 bird species were reported present year-around by Jones, so an estimated minimum of 92 percent of the refuge's total bird diversity is migratory. A recent checklist of 266 species is available from refuge manager, R. R. 1., Box 65, Pingree, ND 58476 (701/285-3341). It is also available on-line: http:// www.npwrc.usgs.gov/resource/birds/chekbird/r6/38.htm

2. Audubon National Wildlife Refuge. 24,700 acres. Situated at the east end of Lake Sakakawea, North Dakota, between Minot and Bismarck. Contains about 13,500 acres administered by the federal government and 11,200 acres supervised by the state. Mostly consists of short-grass prairie and reservoir shoreline, as well as prairie potholes and marshes. At least 78 species are common to abundant during spring, *vs.* 61 species during fall and 6 during winter (Jones, 1990). There are at least 85 nesting birds among the 205 listed for the refuge by Jones. A total of 25 bird species were reported present year-around by Jones, so an estimated minimum of 88 percent of the refuge's total bird diversity is migratory. A recent checklist of 239 species is available from the refuge manager, R.R. 1, Coleharbor, ND 58531 (701/442-5474). It is also available on-line: http://www.npwrc.usgs.gov/resource/birds/chekbird/r6/38.htm

3. Cedar River National Grassland 6,237 ac. Shortgrass and mixed-grass prairie. Address: P.O. Box 390, Lemmon, ND 57638. Ph. 605/374-3592.

4. Chase Lake National Wildlife Refuge. 4,385 acres. Administered through Arrowwood National Wildlife Refuge, this refuge is notable for its enormous nesting colony (10,000–30,000 birds) of American White Pelicans, as well as nesting Double-crested Cormorants, Ring-billed Gulls and California Gulls. Rare species such as Sprague's Pipits, Baird's Sparrows and Piping Plovers also nest here. Public access is seasonally restricted. Managed through Arrowwood N.W.R. For information, contact the Arrowwood refuge manager, 5924 19th St SE, Woodworth, ND 58496 (701-752-4218). East of Chase Lake and near Woodworth is the Woodworth Waterfowl Production Area, and south of Chase Lake near Crystal Springs are two state-owned lakes, Alkali Lake and George Lake. Both are alkaline wetlands with abundant growths of pondweeds and wigeongrass that attract huge numbers of Tundra Swans, Redheads, Canvasbacks, other waterfowl and shorebirds. For information contact the North Dakota Game & Fish Dept., 100 N. Bismarck Expressway, Bismarck, ND 58501 (701/221-6300).

5. Des Lacs National Wildlife Refuge. 18,000 acres. Located one mile west of Kenmare. At least 92 species are common to abundant during spring, *vs.* 69 species during fall and 8 during winter (Jones, 1990). There are at least 145 nesting birds among the 266 listed for the refuge complex by Jones. A total of 25 bird species were reported present year-around by Jones, so an estimated minimum of 91 percent of the refuge complex's total bird diversity is migratory A collective bird checklist of 308 species, including 150 nesters, reported from all three refuges in the "Souris loop" (Des Lacs, J. Clark Salyer and Upper Souris), is available from the manager, Des Lacs Refuge Complex, P.O. Box 578, Crosby, ND 58730 (701/385-4046). It is also available on-line: http://www.npwrc.usgs.gov/resource/birds/chekbird/r6/38.htm

6. Devils Lake Wetland Management District. 221,989 acres This W.M.D. near Devil's Lake has 187 waterfowl production areas totaling 40,113 acres, ten easement refuges totaling 15,891 acres, and 2,521 conservation easements totaling 149,124 acres. It also includes Lake Alice National Wildlife Refuge (12,1794 acres) northwest of Devils Lake, Sullys Hill National Game Preserve (1,5674 acres) southwest of Devils Lake and Kelly's Slough National Wildlife Refuge, located about 12 miles northwest of Grand Forks. Lake Alice attracts some 200,000 Snow Geese in spring, and has nesting colonies of American White Pelican, Double-crested Cormorant, Black-crowned Night-Heron and Franklin's Gull. Stump Lake National Wildlife Refuge (27 acres), south of Lakota, historically supported nesting White-winged Scoters, and some still occasionally stop on migration. The site has long been a major shorebird magnet during spring, and from mid-July to early October. Stump Lake was flooded in 2005, and is now (2012) closed to the public. Kelly's Slough National Wildlife Refuge has a remarkable concentration of migrant waterfowl, shorebirds and wading birds. These wetlands in northeastern North Dakota's moist prairie region are of great importance to migratory shorebirds and waterfowl. Stump Lake and Lake Alice lack bird lists. Kelly's Slough's bird list of 280 species has ten breeding ducks and seven breeding shorebirds. The list is available from the district office and on-line: http://www.npwrc.usgs.gov/resource/birds/chekbird/r6/kellys.htm, Sullys Hill N.G.P. has a

bird list of 267 species. It is available on-line: http://www.npwrc. usgs.gov/resource/birds/chekbird/r6/sullys.htm . For information on Devils Lake W.M.D. contact the district office, 218 SW 4th St, P.O. Box 908, Devils Lake, ND 58301 (701/662-8611).

7. J. Clark Salyer National Wildlife Refuge (formerly Lower Souris N.W.R). 58,700 acres. Located three miles north of Upham. More than 250,000 ducks and 300,000 Snow Geese visit this refuge annually. At least 92 species are common to abundant during spring, *vs.* 69 species during fall and 8 during winter (Jones, 1990). There are at least 145 nesting birds among the 266 listed for the refuge complex by Jones. A total of 25 bird species were reported present year-around by Jones, so an estimated minimum of 91 percent of the refuge complex's total bird diversity is migratory A bird checklist of 308 species, including 150 nesters, collectively reported from all three refuges in the so-called "Souris loop" (Des Lacs, J. Clark Salyer and Upper Souris), is available from the J. Clark Salyer refuge manager, Box 66, Upham, ND 58799 (701/768-2548). It is also available on-line: http://www.npwrc.usgs.gov/resource/birds/chekbird/r6/38.htm

8. Kulm Wetland Management District. 42,352 acres. This district in south-central North Dakota includes several small federal refuges. It also manages 102,000 acres of conservation easements in four counties that are important for migratory waterfowl and shorebirds. No bird list is yet available. Address of W.M.D.: Box E, Kulm, ND 58456 (701/647-2866).

9. Lake Ilo National Wildlife Refuge. 4,043 acres. Located one mile east of Dunn Center. At least 79 species are common to abundant during spring, *vs.* 61 species during fall, and 6 during winter (Jones, 1990). There are at least 83 nesting birds among the 205 listed for the refuge by Jones. A total of 27 bird species were reported present year-around by Jones, so an estimated minimum of 87 percent of the refuge's total bird diversity is migratory. A recent checklist of 226 species is available from the refuge manager, P.O. Box 127, Dunn Center, ND 58626 (701/385-4046). It is also available on-line: http://www.npwrc.usgs.gov/resource/birds/chekbird/r6/38.htm

10. Little Missouri National Grassland, 1,027,852 acres. Shortgrass plains and eroded badlands; the largest area of federally protected shortgrass plains in the nation. Address: Rte. 3., Box 131-B, Dickinson, ND 58601. Ph. 701/225-5152.

11. Long Lake National Wildlife Refuge. 22,300 acres. Situated about four miles southeast of Moffitt, North Dakota, this refuge is mostly prairie grasslands, ravines, fields, trees and shrub plants, and marshy or shallow lake areas. This is one of North Dakota's best stopover sites for migrants, the more common species including Tundra Swans, Canada and Greater White-fronted Geese, Sandhill Cranes, both yellowlegs, a half-dozen "peep" *Calidris* sandpipers, and three gulls. Regular shorebird migrants total at least 16 species. At least 67 species are common to abundant during spring, *vs.* 66 species during fall, and 5 during winter (Jones, 1990). There are at least 78 nesting birds among the 203 listed for the refuge by Jones. A total of 18 bird species were reported present year-around by Jones, so an estimated minimum of 91 percent of the refuge's total bird diversity is migratory. A recent checklist of 289 species is available from the refuge manager, R. R. 1, Moffitt, ND 58560 (701/387-4397). It is also available on-line: http://www.npwrc.usgs.gov/resource/birds/ chekbird/r6/38.htm

12. Lostwood National Wildlife Refuge. 27,647 acres. Located 16 miles southwest of Kenmare. The refuge is a mixture of rolling hills and many small, shallow "pothole" wetlands, some of which are alkaline. The spring and fall migrations include Tundra Swan, Snow and Greater White-fronted geese, and dozens of duck species. At least 13 species of ducks probably nest on the refuge, as well as four grebes, three herons, three terns, two rails, and two gulls. Piping Plovers, Wilson's Phalaropes, California Gulls and American Avocets nest in alkaline areas, and other nesting shorebirds include Spotted Sandpipers, Willets and Marbled Godwits. A recent checklist of 234 species (104 breeders) is available from the refuge manager, Lostwood, N.W.R., R.R. 2, Box 93, Kenmare, ND 58746 (701/848-2722). It is also available on-line: http://www.npwrc.usgs. gov/resource/birds/chekbird/r6/38.htm

13. Sheyenne National Grassland 70,180 ac. Sandhills tallgrass prairie and riverine hardwoods on a sandy glacial-age delta of the Sheyenne River; the largest area of federally owned tallgrass prairie in the U.S.A. Address: Sheyenne Ranger District, P.O. Box 946, 701 Main St., Lisbon, ND 58054. Ph. 701/683-4342.

14. Slade National Wildlife Refuge. 3,000 acres. Situated between Jamestown and Bismarck, North Dakota. Consists of prairie pothole habitat, with many marshes and small lakes. Managed through Long Lake N.W.R. A recent checklist of 202 species, including 88 breeders, is available from the refuge manager of Long Lake N.W.R., R. R. 1, Moffitt, ND 58560 (701/387-4397).

15. Tewaukon National Wildlife Refuge. 8,442 acres. Situated in southeastern North Dakota near Cayuga, and consisting of nearly 8,000 acres of prairie grassland, marshes, and larger water areas. The refuge includes four large wetlands that may attract 100,000 or more Snow Geese. Eared, Western and Pied-billed grebes nest here, as do Willets, Marbled Godwits, and Upland Sandpipers. Swamp, Nelson's and Le Conte's sparrows also nest, and at least two rails. At least 84 species are common to abundant during spring, *vs.* 72 species during fall, and 8 during winter (Jones, 1990). There are at least 98 nesting birds among the 236 listed for the refuge by Jones. A total of 29 bird species were reported present year-around by Jones, so an estimated minimum of 88 percent of the refuge's total bird diversity is migratory. A recent checklist of 236 species is available from the refuge manager, R.R. 1, Cayuga, ND 58103 (701/724-3598). It is also available on-line: http://www.npwrc.usgs.gov/resource/birds/chek-bird/r6/38.htm

16. Upper Souris National Wildlife Refuge. 32,000 acres. Located seven miles north of Foxholm. At least 92 species are common to abundant during spring, *vs.* 69 species during fall and 8 during winter (Jones, 1990). There are at least 145 nesting birds among the 266 listed for the refuge complex by Jones. A total of 25 bird species were reported present year-around by Jones, so an estimated minimum of 91 percent of the refuge complex's total bird

diversity is migratory. A collective bird checklist of 308 species, including 150 nesters, is available from any of the three refuges in the "Souris loop" (Des Lacs, J. Clark Salyer and Upper Souris), or by mail from Upper Souris National Wildlife Refuge, R.R. 1, Foxholm, ND 58738 (701/468-5614), It is also available on-line: http://www.npwrc.usgs.gov/resource/birds/chekbird/r6/38.htm

OKLAHOMA (OK)

1. Black Kettle National Grassland 32,000 ac. Midgrass and tallgrass prairie and upland woods. Partly in Texas. Address: P.O. Box 266, Cheyenne, OK 73628. Ph. 580/497-2143.

2. Optima National Wildlife Refuge. 4,333 acres. Located 15 miles east of Guymon. At least 65 species are common to abundant during spring, *vs.* 63 species during fall, and 21 during winter (Jones, 1990). There are at least 106 nesting birds among the 246 listed for the refuge by Jones. A total of 63 bird species were reported present year-around by Jones, so an estimated minimum of 74 percent of the refuge's total bird diversity is migratory. A recent checklist is available from the refuge manager, c/o Washita NWR, R.R. 1, Box 68, Butler, TX 73625 (405/473-2205). It is also available on-line: http://www.npwrc.usgs.gov/resource/birds/chekbird/r2/40.htm

3. Salt Plains National Wildlife Refuge. 31, 997 acres. This refuge is three miles southeast of Jet, Oklahoma, in Alfalfa County, and is associated with the Salt Plains Reservoir on the Salt Fork of the Arkansas River. Most of the area is covered by the Salt Plains Reservoir, but there is upland forest, rangeland, plus extensive salt flats that provide a unique habitat. Snowy Plovers nesting there are among the refuge's specialties. At least 131 species are common to abundant during spring, *vs.* 110 species during fall, and 43 during winter (Jones, 1990). There are at least 98 nesting birds among the 243 listed for the refuge by Jones. A total of 67 bird species were reported present year-around by Jones, so an estimated minimum of 73 percent of the refuge's total bird di-

versity is migratory. Large flocks of American White Pelicans use the site during migration, and Whooping Cranes are regular transients. A recent checklist of 296 species is available from the refuge manager, Rte. 1., Box 76. Jet, OK 73749 (405/626-4794). It is also available on-line: http://www.npwrc.usgs.gov/resource/birds/chekbird/r2/40.htm

4. Sequoyah National Wildlife Refuge. 20,800 acres. This refuge is in east-central Oklahoma, around the western part of the Robert S. Kerr Reservoir, and about half of the acreage is water. Most of the rest is shoreline or river bottomland, with many ponds and sloughs. At least 78 species are common to abundant during spring, *vs.* 71 species during fall, and 57 during winter (Jones, 1990). There are at least 96 nesting birds among the 250 listed for the refuge by Jones. A total of 56 bird species were reported present year-around by Jones, so an estimated minimum of 78 percent of the refuge's total bird diversity is migratory. A recent checklist of 256 species seen since 1970 is available from the refuge manager, Box 398, Sallisaw, OK 74955 (918/773-5251). It is also available on-line: http://www.npwrc.usgs.gov/resource/birds/chekbird/r2/40.htm

5. Tishomingo National Wildlife Refuge. 16,600 acres. This refuge is six miles southeast of Tishomingo, on Lake Texoma in eastern Oklahoma. It includes about 4,000 acres of reservoir, as well as marshes, cropland and grassland. At least 113 species are common to abundant during spring, *vs.* 94 species during fall, and 65 during winter (Jones, 1990). There are at least 81 nesting birds among the 243 listed for the refuge by Jones. A total of 72 bird species were reported present year-around by Jones, so an estimated minimum of 70 percent of the refuge's total bird diversity is migratory. A recent checklist of 252 species is available from the refuge manager, P.O. Box 248, Tishomingo, OK 73460 (405/371-2402). It is also available on-line: http://www.npwrc.usgs.gov/resource/birds/chekbird/r2/40.htm

6. Washita National Wildlife Refuge. 8,200 acres. Located 15 miles northwest of Clinton. At least 58 species are common to abun-

dant during spring, *vs.* 54 species during fall, and 40 during winter (Jones, 1990). There are at least 67 nesting birds among the 220 listed for the refuge by Jones. A total of 59 bird species were reported present year-around by Jones, so an estimated minimum of 73 percent of the refuge's total bird diversity is migratory. A recent checklist of 229 species is available from the refuge manager, Information is available from the refuge manager, R.R. 1, Box 68, Butler, TX 73625 (405/473/2205). It is also available on-line: http://www.npwrc.usgs.gov/resource/birds/chekbird/r2/40.htm

SOUTH DAKOTA (SD)

1. Black Hills National Forest. 1,235.453 acres. This national forest includes the highest elevation of any point in the northern Great Plains (7,242 feet), and supports many montane forest species associated with the Rocky Mountains of Wyoming. Of the 131 regularly occurring species, 33 are permanent residents, 60 are summer residents, 27 are transients, and 11 are overwintering migrants (Pettingill and Whitney, 1965) . Thus, at least 75 percent of the regional birds are migratory. Many of the summer residents are Neotropical migrants, such as the White-throated Swift, Western Wood-Pewee, Hammond's, Dusky and Cordilleran flycatchers, Say's Phoebe, Eastern and Western kingbirds, Ovenbird, Virginia's, Yellow, Yellow-rumped and Mac-Gillivray's warblers, Common Yellowthroat, Yellow-breasted Chat, American Redstart, Plumbeous, Red-eyed and Warbling vireos, Western Tanager and Black-headed Grosbeak. Of these species, 11 are distinctly western in geographic orientation. A bird checklist of 190 species may be obtained from the Forest Headquarters: P.O. Box 792, Custer, SD 57730, Custer, SD 57730 (605/763/2251).

2. Buffalo Gap National Grassland 591,771 ac. Shortgrass and midgrass prairie. Bird list available (197 spp.). Address: P.O. Box 425, U.S.F.S., 708 Main St., Wall, SD 57790, or U.S.F.S., 209 N. River, Hot Springs, SD. 57747. Ph. 605/279-2125 or 605/745-4107.

3. Fort Pierre National Grassland 115,996 ac. Shortgrass and mid-grass prairie. Address: P.O. Box 417, 124 S. Euclid Ave., Pierre, SD 57501. Ph. 605/224-5517.

4. Grand River National Grassland 156,000 ac. Shortgrass and mid-grass prairie. Address: P.O. Box 390, Lemmon, SD 57638. Ph. 605/374-3592.

5. Huron Wetland Management District. 87,5000 acres. This wetland district manages w 62 waterfowl production areas (W.P.A.s) in eight central counties: Beadle, Buffalo, Hand, Hughes, Hyde, Jeraud, Sanborn and Sully. Three of the notable waterfowl production areas in the district are: LeClair W.P.A., 13 miles northwest of Iroquois, Bauer W.P.A., 13 miles east of Huron, and Campbell W.P.A., 15 miles southeast of Miller. District office address: Room 308 Federal Bldg., 200 Fourth St. SW, Huron SD 57360 (Ph. 805-353-5894). No bird list is yet available.

6. Lacreek National Wildlife Refuge. 16,250 acres. Situated about 15 miles southeast of Martin, South Dakota. This refuge at the northern edge of the Nebraska Sandhills consists of extensive marshes and shallow lakes in the Lake Creek valley of the White River's South Fork. Trumpeter Swans are present during most seasons, but may move south into the Nebraska Sandhills during winter. The refuge produces 15–20 Trumpeter Swans each summer, as well as up to 6,000 ducks and 800 Canada Geese. It also has one of the state's three nesting colonies of American White Pelicans. Consists of extensive marshes and shallow lakes in the valley of the South Fork of the White River, just north of the Nebraska Sandhills. At least 77 species are common to abundant during spring, vs. 53 species during fall, and 15 during winter (Jones, 1990). There are at least 93 nesting birds among the 213 listed for the refuge by Jones. A total of 56 bird species were reported present year-around by Jones, so an estimated minimum of 74 percent of the refuge's total bird diversity is migratory. A recent checklist of 273 species (46 breeders), seen on the refuge since 1936 is available from the refuge manager, Martin, SD 57551 (605/685-6508). It is also available on-line: http://www.npwrc.usgs.gov/resource/birds/chekbird/r6/46.htm

7. Lake Andes National Wildlife Refuge. 5,450 acres. This refuge is located north of Fort Randall Dam in Charles Mix County. Lake Andes N.W.R. includes Lake Andes, a shallow Pleistocene glacial lake and marsh, occupying 4,700 acres of open water and marsh. The state-controlled Lake Andes Wetland Management District encompasses 34,682 acres of wetlands and grassland easements over a 13-county area of southeastern South Dakota, including Aurora, Bon Homme, Brule, Charles Mix, Clay, Davison, Douglas, Hanson, Hutchinson, Lincoln, Turner, Union and Yankton counties. Lake Andes N.W.R. is situated north of Fort Randall Dam in southeastern South Dakota. It consists of 5,450 acres, around the Lake Andes marsh. The nearby Karl E. Mundt National Wildlife Refuge is located below Fort Randall Dam. It occupies less than 1,000 acres, but is an important wintering area for Bald Eagles. At least 113 species are common to abundant at the Lake Andes refuge complex during spring, *vs.* 94 species during fall, and 17 during winter (Jones, 1990). There are at least 85 nesting birds among the 214 listed for the area by Jones. A total of 45 bird species were reported present year-around by Jones, so an estimated minimum of 79 percent of the area's total bird diversity is migratory. A recent checklist for the complex (including both refuges and the wetland management district), has 213 species, with 97 wetland species (37 breeders), and is available from the refuge manager, Box 391, Lake Andes, SD 57356 (605/487-7603). It is also available on-line: http://www.npwrc. usgs.gov/resource/birds/chekbird/r6/46.htm

8. Madison Wetland Management District. 52,000 acres. This wetland district in the heart of the glaciated pothole region of eastern South Dakota manages wetlands ranging in size from 40–400 acres, and includes over 36,000 acres preserved in waterfowl production areas. It includes eight east-central counties: Brookings, Deuel, Hamlin, Kingsbury, Lake, McCook, Minnehaha, Moody, Sanborn and Sully. A recent checklist containing 297 species, including 96 wetland species (42 breeders) that have been observed in the district, is available from the Madison Wetland District Office, Box 48, Madison, SD 57042 (Ph. 605-256-2974). It is also available on-line: http://www.npwrc.usgs.gov/resource/birds/ chekbird/r6/madison.htm.

9. Pocasse National Wildlife Refuge. 2,540 acres. Located just north of Pollock, Brown County, off U.S. Hwy. 83, and bordering the east side of the Missouri River. Mostly marshes and open water (1,045 acres of wetland), this refuge is an important stopover area for migrating Sandhill and Whooping cranes, as well as for waterfowl. No bird list for Pocasse N.W.R. is yet available. Administered from Sand Lake N.W.R. (Ph. 605-885-6320). General information may be obtained from the refuge manager, c/o Sand Lake NWR, R.R. 1, Box 253, Columbia, SD 57433 (Ph. 605-885-6320).

10. Samuel H. Ordway Jr. Memorial Prairie Preserve 7,800 ac. Tallgrass prairies and wetlands, the largest tallgrass prairie in South Dakota, and an important grassland/wetland preserve for migrating birds. Owned by The Nature Conservancy. Address: 35333 115th Street, Leola, SD 57456 Ph. 605/439-3475.

11. Sand Lake National Wildlife Refuge. 45,000 acres. This refuge is located 25 miles northeast of Aberdeen, Brown County, in the James River Valley, and was part of the shoreline of glacial Lake Dakota until about 10,000 years ago. It consists of more than 21,000 acres of marshes, grasslands, shallow impoundments, shelterbelts and fields. Sand Lake N.W.R. has the world's largest nesting colony of Franklin's Gulls, and has been identified as a Ramsar wetland of international importance. It attracts hundreds of thousands of Snow Geese and other waterfowl during migration, and attracts four breeding grebe species (Eared, Western, Clark's and Pied-billed), as well as nesting Canvasbacks, Redheads, Lesser Scaup and Ruddy Ducks. Notable nesting shorebirds include the Marbled Godwit, Wilson's Phalarope and three terns (Common, Forster's and Black). The refuge also administers Pocasse National Wildlife Refuge (see above), and is located among more than 150,000 acres of regional state-owned wildlife management areas in the glaciated pothole region of northeastern South Dakota. The nearby Sand Lake Wetland Management District is the largest wetland management district in the country, encompassing 9,000 square miles. The district includes ten of South Dakota's north-central counties: Brown, Campbell, Corson, Dewey, Edwards, Faulk, McPherson, Potter, Spink and Wal-

worth. It contains 45,000 acres of land under federal protection, involving 162 waterfowl production areas, and includes an additional 550,000 acres protected by conservation easements. At least 91 species are common to abundant during spring, *vs.* 60 species during fall, and 10 during winter (Jones, 1990). There are at least 111 nesting birds among the 239 listed by Jones. A total of 49 bird species were reported present year-around by Jones, so an estimated minimum of 80 percent of the refuge's total bird diversity is migratory. A recent checklist for Sand Lake N.W.R. totals 263 species, including 106 wetland species (of which 55 are breeders, the largest number for any regional site), and is available on-line or from the refuge manager, R.R. 1, 399650 Sand Lake Drive, Columbia, SD 5433 (Ph. 605-885-6320). It is also available on-line: http://www.npwrc.usgs.gov/resource/birds/chekbird/r6/46.htm

12. Waubay National Wildlife Refuge. 4,600 acres. This refuge is situated eight miles north of Waubay, Day County, in the glaciated till region of northeastern South Dakota, the heart of South Dakota's pothole country. It contains nearly 5,000 acres of marshlands, lakes, grasslands, brush, and woodlands. It has all the species of nesting grebes as those mentioned for Sand Lake National Wildlife Refuge, plus Horned and occasional Red-necked grebes. It also attracts the same nesting species of diving ducks, shorebirds and terns. At least 106 species are common to abundant at Waubay N.W.R. during spring, *vs.* 105 species during fall, and 7 during winter (Jones, 1990). There are at least 109 nesting birds among the 244 listed for the refuge by Jones. A total of 29 bird species were reported present year-around by Jones, so an estimated minimum of 88 percent of the refuge's total bird diversity is migratory. A recent bird checklist containing 244 species, including 103 wetland species (52 breeders), is available from the refuge manager, R.R. 1, Waubay, SD 57273 (605/947-4695). It is also available on-line: http://www.npwrc.usgs.gov/resource/birds/chekbird/r6/46.htm

13. Waubay Wetland District, This district includes over 300 waterfowl production areas in six wetland-rich northeastern counties: Clark, Codington, Day, Grant, Marshall and Roberts. The district

is headquartered at Waubay, South Dakota, in the glaciated Coteau Hills of northeastern South Dakota, it contains nearly 5,000 acres of marshlands, lakes, grasslands, brushy areas, and oak timber. The bird list for Waubay N.W.R. (see above) is probably applicable here. Address of district wetland office: 44401 134A St., Waubay, SD 57273 (Ph. 605-947-4521).

TEXAS (TX)

1. Anahuac Wildlife Refuge. 24,356 acres. Located about 10 miles southeast of Anahuac. At least 100 species are common to abundant at the refuge during spring, *vs.* 100 species during fall (Jones, 1990). There are at least 40 nesting birds among the 252 listed for the refuge by Jones. A total of 49 bird species were reported present year-around by Jones, so an estimated minimum of 81 percent of the refuge's total bird diversity is migratory. A recent checklist of 279 species is available from the refuge manager, Trinity St. & Washington Ave., P.O. Box 278, Anahuac, TX 77514 (409/267-3337). It is also available on-line: http://www.npwrc. usgs.gov/resource/birds/chekbird/r2/48.htm. Anahuac N.W.R. also administers McFaddin and Texas Point National Wildlife refuges (total area 63,835 acres), which provide important winter estuarine habitats for vast numbers of wintering ducks and geese. Up to 100,000 ducks of 23 species winter at Texas Point, as well as large numbers of geese (Snow, Canada and Greater White-fronted). It is located 17 miles south of Port Arthur and has no vehicular roads, so public access to both is very limited. Progressively farther to the west are Sabine Woods Sanctuary, Sea Rim State Park and McFaddin National Wildlife Refuge. Sabine Woods Sanctuary is owned by the Texas Ornithological Society, and its spring bird migration is said to nearly equal the better-known High Island. It is four miles west of Sabine Pass on State Hwy. 87, and might be easily missed as it is poorly marked, if at all. McFaddin N.W.R. has 12 miles of beach and eight miles of interior roads. Information is available at the refuge's field headquarters, P,O, Box 609, Sabine Pass, TX (409/971-2909). A few miles west of Anahuac N.W.R. is Candy Cain Abshier State Wild-

life Management Area, which is best known as a viewing point
for fall raptor migrations. From early September to mid-October
thousands of Broad-winged hawks, as well as hundreds of other
buteos, accipiters, falcons, eagles, kites and vultures can be seen.
In one year, fall totals at Smith Point exceeded 12,000 birds and
included 7,766 Broad-winged Hawks, 3,675 Mississippi Kites, 354
Sharp-shinned Hawks, 246 Cooper's Hawks, 166 American Kestrels, 97 Swallow-tailed Kites, 51 Swainson's Hawks, 51 Northern
Harriers, 43 Merlins, 39 Peregrine Falcons, 31 Ospreys, 23 Red-
shouldered Hawks, 10 Red-tailed Hawks, 6 White-tailed Kites, 5
Crested Caracaras, 5 White-tailed Hawks and 1 Bald Eagle.

2. Aransas National Wildlife Refuge. 54,829 acres. Located seven
miles south of Austwell. There are 252 species listed by Jones
(1990) for the refuge, of which at least 92 species are common at
the refuge during spring, *vs.* 76 species during summer, 74 during fall, and 69 during winter (Jones, 1990). A total of 79 bird
species were reported present year-around by Jones, so an estimated minimum of 69 percent of the refuge's total bird diversity
is migratory. Essentially the entire population of the highly endangered Whooping Cranes that breeds in Alberta's Wood Buffalo National Park winters here, the early arrivals usually ap-
pearing by late October and all leaving by early April. As of 2012
this slowly recovering population numbered nearly 300 birds. A
recent checklist of 392 species is available from the refuge manager, P.O. Box 100, Austwell, TX 77950 (512/286-3559). It is also
available on-line: http://www.npwrc.usgs.gov/resource/birds/
chekbird/r2/48.htm

3. Big Bend National Park and Big Bend Ranch State Park. Big Bend
National Park (801,163 acres) is largely desert, but the bordering
Rio Grande River provides a limited water source. The park's bird
checklist is notable for its huge number of wintering and nesting
birds, with at least 388 documented species, plus 59 hypothetical
species. Some 50 species of warblers have been reported, of which
43 are migrants or accidentals. The warblers include one nesting
species, the Colima Warbler that is endemic to the park. The adjoining Big Bend Ranch State Park is smaller (269,714 acres) and

has more water in the form of springs, tanks and waterfalls, as well as associated riparian habitats. It has no bird checklist, but the national park checklist is probably applicable. Address of Big Bend National Park: Headquarters Dr., Big Bend National Park, TX 79834 (432/ 477-2251). Address of Big Bend Ranch State Park: P.O. Box 2319 Presidio, TX 79845 ((432) 358-4444).

4. Big Thicket National Preserve. 844,500 acres. Consists of several parcels north of Beaumont. This national preserve includes eight land units and four river or stream corridor units. There is a great diversity of ecosystems, such as pine-hardwood forest, longleaf pine-savanna forest, beech-magnolia-loblolly pine forest, bald cypress–tupelo swamps, bog wetlands, riparian hardwoods and others. A bird checklist with 235 species is available at preserve headquarters. Many migratory species nest here, including such Neotropical migrants as at least 13 warblers and three vireos. Many more northerly nesting migrants winter here. For information contact the preserve superintendent, 3785 Milam St., Beaumont, TX 77770 (409/839-2689).

5. Black Kettle National Grassland. 31,000 acres. Upland mixed prairie and hardwoods. See Oklahoma account for description.

6. Bolivar Flats Shorebird Sanctuary. 615 acres. This area of tidal mudflats, salt marsh, beach and uplands is located at the south end of the Bolivar Peninsula at the head of Galveston Bay, in the unincorporated community of Port Bolivar. It is managed by the Houston Audubon Society, and provides resting, feeding and breeding habitats for hundreds of thousands of shorebirds and other water-dependent birds annually. It has been designated as a nationally recognized Important Bird Area, and a Wetland of International Importance by the Western Hemisphere Shorebird Reserve Network, in part because of the large numbers of Piping Plovers and Snowy Plovers that use this area as a migratory stopover and wintering site. It is also heavily used by thousands of wading birds for resting or foraging, such as Reddish Egrets and Roseate Spoonbills. Various herons and egrets are permanent residents, and a wide variety of gulls

and terns are often present. For information contact the Houston Audubon Society, 440 Wilchester Boulevard, Houston, TX 77079 (713/932-1639).

7. Brazoria National Wildlife Refuge. 10,407 acres. Located about ten miles northeast of Freeport. At least 98 species are common to abundant at the refuge during spring, *vs.* 86 species during fall, and 81 during winter (Jones, 1990). Wintering birds include large numbers of Snow, Ross's, Canada and Greater White-fronted geese; up to 100,000 geese and 80,000 ducks of 24 species have been seen. There are at least 71 nesting birds among the 272 listed for the refuge by Jones. A total of 99 bird species were reported present year-around by Jones, so an estimated minimum of 64 percent of the refuge's total bird diversity is migratory. The refuge is open to the public on the first full weekend of each month, and on the third weekend from November through April, with limited access during the week. The nearby Big Boggy N.W.R. has similar habitats, but is usually not open to the public. A recent Brazoria checklist with 301 species is available from the refuge manager, 1218 N. Velasco, P.O. Drawer 1088, Angleton, TX 77515 (409/859-6062). It is also available on-line: http://www.npwrc.usgs.gov/resource/birds/chekbird/r2/48.htm

8. Buffalo Lake National Wildlife Refuge. 7,700 acres. Situated 30 miles southwest of Amarillo, Texas. This refuge includes about 1,000 acres of surface water resulting from the impoundment of Tierra Blanca Creek, as well as adjoining grasslands. At least 28 species are common to abundant at the refuge during spring, *vs.* 19 species during fall, and 20 during winter (Jones, 1990). There are at least 42 nesting birds among the 246 listed for the refuge by Jones. At least 36 shorebirds have been documented on the refuge, and 17 sparrows also have occurred, including all four longspurs. The refuge's alkaline wetlands are a major wintering area for lesser Sandhill Cranes. At least 28 warblers are on the refuge list, of which at least five nest. A total of 42 bird species were reported present year-around by Jones, so an estimated minimum of 83 percent of the refuge's total bird diversity is migratory. A recent checklist contains 344 species and is available from the ref-

uge manager, P.O. Box 228, Umbarger, TX 79091 (806/499-3352). It is also available on-line: http://www.npwrc.usgs.gov/resource/birds/chekbird/r2/48.htm

9. Caddo National Grassland 17,796 ac. Grasslands and hardwoods. Address: P.O. Box 507, Decatur, TX 76234. Ph. 817/627-5475.

10. Guadalupe Mountains National Park. Area 86,415 acres. This park on the New Mexico border contains the highest mountains in Texas, and has an elevation range of over 8,000 feet. As such, it has a wide variety of climates and habitats, ranging from subalpine pine forests and grassy meadows to wooded deciduous canyons and arid scrub desert. Over 300 mostly bird species have been reported, including many western-oriented Neotropical migrants such as the White-throated Swift, Western Wood-Pewee, Cordilleran Flycatcher, Violet-green Swallow, Virginia's Warbler, Gray Vireo and Scott's Oriole. A bird checklist list is available at the Headquarters Visitors Center, HC 60, Box 400, Salt Flat, TX 79847-9400 (915/828-3251).

11. Hagerman National Wildlife Refuge. 11,320 acres. Situated 15 miles northwest of Sherman, Texas, around Lake Texoma, on the Texas-Oklahoma border. This refuge is associated with Lake Texoma and includes habitats similar to those in the Tishomingo National Wildlife Refuge in Oklahoma. The winter refuge list includes five goose species, 21 ducks, and 36 species of shorebirds. At least 28 warblers occur regularly, six of which remain to nest. At least 63 species are common to abundant at the refuge during spring, *vs.* 67 species during fall, and 46 during winter (Jones, 1990). There are at least 86 nesting birds among the 272 listed for the refuge by Jones. A total of 76 bird species were reported present year-around by Jones, so an estimated minimum of 72 percent of the refuge's total bird diversity is migratory. A recent checklist containing 316 species is available from the refuge manager, Rte. 3, Box 123, Sherman, TX 75090 (214/786-3826). It is also available on-line: http://www.npwrc.usgs.gov/resource/birds/chekbird/r2/48.htm

12. High Island Audubon sanctuaries. The town of High Island is 30 miles northeast of Galveston, and is slightly higher in elevation than surrounding areas of coastal prairie. There are several groves of live oaks in town, which are perfect resting points for Neotropic migrants that have just crossed the Gulf of Mexico. There are four sanctuaries at High Island, owned and managed by the Houston Audubon Society. These are Smith Oaks (142 acres), Louis B. Smith Woods (50+ acres), Eubank Woods (9.5 acres) and S. E. Gast Red Bay Sanctuary (8.8 acres). Great numbers and a wide variety of warblers, vireos, orioles, buntings, and other migrants can be seen here from mid-April to mid-May, especially after a cold front accompanied by rain has passed. For information contact Houston Audubon Society, 440 Wilchester Boulevard, Houston, TX 77079 (713/932-1639).

13. Laguna Atascosa National Wildlife Refuge. 45,187 acres. Located 25 miles east of Harlingen. At least 97 species are common to abundant at the refuge during spring, *vs.* 95 species during fall, and 67 during winter (Jones, 1990). There are at least 90 nesting birds among the 329 listed for the refuge by Jones. A total of 122 bird species were reported present year-around by Jones, so an estimated minimum of 63 percent of the refuge's total bird diversity is migratory. Seven vireos and 35 warblers have been reported; two of the warblers remain to nest. There is a small re-introduced population of Aplomado Falcons, and great numbers of wintering ducks, especially Redheads. A recent checklist containing 369 species is available from the refuge manager, P.O. Box 450, Rio Hondo, TX 78583 (512/748-3607). It is also available on-line: http://www.npwrc.usgs.gov/resource/birds/chekbird/r2/48.htm

14. Lower Rio Grande National Wildlife Refuge. Located seven miles south of Alamo, sharing a headquarters with Santa Ana National Wildlife Refuge. This still-developing National Wildlife Refuge follows the lower Rio Grande River in Starr, Hidalgo, Cameron and Willacy Counties, and may eventually have as many as 132,500 acres. The bird list is remarkably large (the largest of any refuge in the Great Plains), and contains 402 species, including

136 nesters, suggesting that a minimum of 66 percent of the refuge's total bird diversity is migratory. The list contains 39 shorebirds, 24 New World flycatchers, and 40 warblers nearly all of which are Neotropical migrants. For information contact the refuge manager at 3325 Green Jay Road, Alamo, Texas 78516 (956/784-7500).

15. Lyndon B. Johnson National Grassland 20,320 acres. Grasslands and hardwoods. Address: P.O. Box 507, Decatur, TX 76234, Ph. 817/627-5475..

16. McFaddin National Wildlife Refuge. 55,000 acres. This refuge protects the largest remaining freshwater marsh in Texas, and is adjacent to thousands of acres of variably brackish marshland. Tens of thousands of geese and ducks winter here, and the increasingly rare Mottled Duck is a year-around resident. The refuge is located about 15 miles south of Port Arthur, P. O. Box 358, 7950 S. Gulfway Dr., Sabine Pass, TX 77655 (409/971-2909). http://www.fws.gov/southwest/refuges/texas/mcfaddin/

17. Muleshoe National Wildlife Refuge. 5,809 acres. Situated 20 miles south of Muleshoe. The refuge contains 5,800 acres of lakes, marshes, short-grass plains, and other minor habitats. The sink-type lakes provide the most important wintering habitat in North America for lesser Sandhill Cranes, which sometimes number over 100,000, and during some winters have reached as many as 250,000. At least 78 species are common to abundant at the refuge during spring, *vs.* 82 species during fall, and 49 during winter (Jones, 1990). There are at least 59 nesting birds among the 243 listed for the refuge by Jones. A total of 65 bird species were reported present year-around by Jones, so an estimated minimum of 73 percent of the refuge's total bird diversity is migratory. The bird checklist includes 282 species and is available from the refuge manager, P.O. Box 549, Muleshoe, TX 79347 (806/946/3341). It is also available on-line: http://www.npwrc.usgs.gov/resource/birds/chekbird/r2/48.htm

18. Rita Blanca National Grassland 77,413 ac. Shortgrass and mid-grass prairies, partly in New Mexico. Address: Box 38, Texline, TX 79807. Ph. 806/362-4254.

19. Rockport and Connie Hagar Wildlife Sanctuary. This town on the central Gulf Coast is a birding "hotspot," largely because of the large numbers of Ruby-throated Hummingbirds that pass through during fall migration, which are the basis for a birding celebration each September. At least 413 bird species have been tallied in the central Gulf Coast region around Rockport. A part of Little Bay, from northern Rockport to the mouth of Fulton Harbor, has been designated as the Connie Hagar Wildlife Sanctuary. For information, contact the Rockport Chamber of Commerce, 404 Broadway, Rockport TX 78382 (512/729-6445), or the Hummingbird Lodge and Education Center, HCO 1, Box 245. FM 1781, Rockport, TX 78382 (512/729-7555).

20. San Bernard National Wildlife Refuge. 27,414 acres. Located about 10 miles south of Lake Jackson, and 12 miles west of Freeport. San Bernard, Brazoria ad Big Boggy refuges collectively winter many migratory waterfowl, especially Snow Geese, with Greater White-fronted, Canada and Ross's geese present in smaller numbers. A recent checklist for the three-refuge complex contains 301 species and is available from the Brazoria refuge manager, 1218 N. Velasco, P.O. Drawer 1088, Angleton, TX 77515 (409/849-6062). It is also available on-line: http://www.npwrc.usgs.gov/resource/birds/chekbird/r2/48.htm

21. Santa Ana National Wildlife Refuge. 2,088 acres. Located seven miles south of Alamo. At least 61 species are common to abundant at the refuge during spring, vs. 53 species during fall, and 41 during winter (Jones, 1990). There are at least 87 nesting birds among the 334 listed for the refuge by Jones. A total of 92 bird species were reported present year-around by Jones, so an estimated minimum of 73 percent of the refuge's total bird diversity is migratory. The refuge list has a remarkable 43 warbler species, most of them migratory, and many other migrants stop here. During migration thousands of migrating Broad-winged Hawks pass overhead. A recent checklist includes 396 species

and is available from the refuge manager, Rte. 2, Box 202A, Alamo, TX 78516 (512/787-3079). It is also available on-line: http://www.npwrc.usgs.gov/resource/birds/chekbird/r2/48.htm

22. Texas Point National Wildlife Refuge. 8,900 acres. Consists of fresh water to salty marshland, with some wooded uplands and prairie ridges. This refuge is located 12 miles east of McFaddin NWR, along Highway 87 and near the Louisiana border. Like nearby McFaddin NWR,, it is an important wintering area for wintering geese and ducks. It is managed out of McFaddin, and uses the same address, phone number and website. Not far to the east, along the western Louisiana coast, are Sabine NWR (142,000 acres), and Lacassine NWR (31,700 acres). Both are similar in their coastal habitats to Texas Point NWR and McFaddin NWR, and have large diversities of breeding and wintering birds,. However, since Louisiana is considered to fall within the boundaries of Mississippi Flyway, these refuges are not included in this work.

23. Welder Wildlife Foundation and Hazel Bazemore County Park. Welder Wildlife Foundation is a privately owned nature preserve of 7,800 acres, It is located eight miles north of Sinton and includes costal plains and upland habitats. The site's bird list includes 372 species, 96 of which nest or have nested on the preserve, and includes 41 shorebirds, 24 hawks, and 41 warblers. For information, contact the Foundation at P.O. Drawer 1400, Sinton, TX 78387 (512/364-2643). To the south of Sinton, near Corpus Christi, is Hazel Bazemore County Park, operated by the Nueces County Parks and Recreation Department. This 78-acre park is located on the Nueces River. During one fall hawk-watchers at Hazel Bazemore Park tallied over 690,000 individuals of 24 raptor species. These included 678,204 Broad-winged Hawks, 9,735 Mississippi Kites, 467 American Kestrels, 460 Cooper's Hawks, 367 Sharp-shinned Hawks, 281 Swainson's Hawks, 154 Ospreys, 125 Peregrine Falcons, 92 Red-tailed Hawks, 48 Merlins, 34 Northern Harriers, 21 Swallow-tailed Kites, 17 Red-shouldered Hawks, 15 Crested Caracaras, 11 Prairie Falcons, 6 Zone-tailed Hawks, 3 White-tailed Hawks, 3 Harris's Hawks, 2 Ferruginous Hawks, 1 Bald Eagle, 1 Aplomado Falcon and 1 Hook-billed Kite.

WYOMING (Eastern) (WY)

1. Hutton Lake National Wildlife Refuge. 1,968 acres. Located about
   ten miles south of Laramie, at about 7,000 feet elevation. This
   small refuge's arid grassland supports a good population of
   white-tailed prairie dogs (and attract raptors such as Ferruginous
   Hawks and Golden Eagles), Mountain Plovers, McCown's and
   Chestnut-collared longspurs, and grassland sparrows such as
   Chipping, Lark and Lark Bunting. Several small lakes are present that vary in size and depth, with the one nearest the refuge
   entrance the largest and deepest. It often has a variety of migrant
   dabbling and diving ducks, plus grebes (Eared and Western). The
   most distant and least accessible wetlands include a marshy area
   that attracts many migrant shorebirds and dabbling ducks. Managed from Arapahoe NWR, Colorado, the refuge roads are barely
   passable rutted trails, and lacks any facilities. A refuge checklist is not available, but general information is available from the
   refuge manager, c/o Arapahoe NWR, P.O. Box 457, Walden, CO
   80480 (303/482-5155). In the same general region (off Hwy. 230)
   is a flood-control reservoir called Hattie Lake. Depending on the
   amount of water present, it may attracts large numbers of diving
   ducks such as Redheads, and many shorebirds, gulls and American White Pelicans.

2. Lake DeSmet and the Bighorn Mountains. Lake DeSmet is a historic lake between Buffalo and Story, near the eastern slope of the
   Bighorns, and one of the few natural lakes in eastern Wyoming.
   Clear Creek provides the lake's water supply, and for many
   years the Texaco Oil Co. controlled its levels. In the 1970's the
   lake levels were raised 40 feet, covering the mudflats and sandbars that had been used by shorebirds and wading birds. Never-
   theless, it is a still magnet for migrating loons, diving ducks, and
   gulls. Western Grebes stage here is substantial numbers during
   migration, and migrant Common, Red-throated, Pacific and Yellow-billed loons have all been seen here. The birds of the nearby
   Bighorn Mountain region have been surveyed by Canterbury
   and Johnsgard (in prep.), who documented over 320 species. The
   region's long-distance migrants include 34 waterfowl, 34 shore-

birds and 23 warblers; many of the breeders have Rocky Mountain affinities.

3. Ocean Lake Wildlife Habitat Management Area. 12,750 acres. Located 17 miles northwest of Riverton, off Hwy. 134. About half of this area consists of a large, shallow lake with marshy edges that attract many migrating ducks and grebes. The site also attracts Sandhill Cranes, about 400 of which stage here during spring and fall, and some remain to nest. Up to 3,000 geese and 10,000 ducks stop here during migration. Whooping Cranes also stop here rarely. There are nesting colonies of Yellow-headed Blackbirds, Western and Clark's grebes, Forster's Terns and Double-crested Cormorants. Owned by the Wyoming Dept. of Game and Fish; for information contact the state headquarters at 5400 Bishop Rd, Cheyenne, WY 2006 (207/777-4600).

4. Table Mountain Wildlife Habitat Management Unit. 1,716 acres. Located about 15 miles southeast of Torrington. This state-owned area is a major migration stopover point for waterfowl in spring; during fall it is open to controlled waterfowl hunting. The marsh attracts thousands of Snow Geese and hundreds of American White Pelicans, as well as Canada, Ross's and Greater White-fronted geese, plus dozens of duck species. Migrant shorebirds include many "peep" (*Calidris*) sandpipers, especially Stilt Sandpipers. Snowy Plovers, American Avocets and Wilson's Phalaropes nest here. Other migratory water birds using the area and probably nesting are the American Bittern, Great Blue Heron, Western Grebe and White-faced Ibis. About ten miles to the west (and about five miles south of Yoder) is Bump Sullivan Reservoir and nearby Springer Lake, the later owned by the Wyoming Dept. of Game and Fish. Bump Sullivan Reservoir's migrant birds are much like those of Table Mountain, but the reservoir lacks marshy habitats. Springer Lake is an alkaline wetland that is notable for its migrant shorebirds, as well as American White Pelicans, Double-crested Cormorants, Canada and Snow geese, Sandhill Cranes and grebes. For information contact the Wyoming Dept. of Game and Fish; headquarters at 5400 Bishop Rd, Cheyenne, WY 2006 (207/777-4600).

5. Thunder Basin National Grassland 572,319 ac. Shortgrass and shrub steppe plains, the largest area of protected prairie in Wyoming. Bird list of 231 species. The Mountain Plover, Upland Sandpiper, Long-billed Curlew and both prairie longspurs are common. Address: 809 9th St., Douglas, WY 82633. Ph. 307/358-4690.

6. Yant's Puddle. 2,200 acres. This area, just north of Casper, has a history of being one of eastern Wyoming's best birding areas, in spite of its unsavory history. Acquired by Standard Oil of Indiana in the 1950's, and now owned by BP (formerly known as British Petroleum), it served as a repository for effluent wastewater. Over time the resulting wetland (Soda Lake) expanded to more than two square miles in area, and was sufficiently alkaline to support heavy growths of algae and alkaline-adapted invertebrates. A colony of more than 2,000 California gulls eventually developed, as well as colonies of nesting Caspian Terns, Black-crowned Night-Herons and Black-necked Stilts. Snowy Egrets, American Avocets and Wilson's Phalaropes also began nesting. The site also attracted many rare migratory shorebirds, gulls and jaegers. After the refinery closed, the water supply disappeared, with the wetland's future dependent on regular pumping of river water. The wetland has since become ever smaller and excessively saline, killing the plants and invertebrates, and making it a lethal brew for any birds that swim in it. At the time of writing (2012), there was very little water left, and its current manager Atlantic Richfield (a subsidiary of BP) has not taken action to provide any more water, Thus the fate of this once-flourishing and valuable wetland is in great doubt (see article by Ted Williams in *Audubon Magazine*, January–February, 2012).

Canada

Note: For additional information on Canadian sites, contact Environment Canada – Canadian Wildlife Service, Twin Atria Bldg., 2nd floor, 4999 98th Ave., Edmonton, Alberta, Canada   T6B 2X3 (401/468-8075). National wildlife areas are federally owned areas that may have limited access. Visiting a Migratory Bird Sanctuary requires advance permission from the Canadian Wildlife Service, and may require up to 45 days of advance notice. Ramsar sites are designated as wetlands of international importance.

ALBERTA (AB)

Note: For further information on Alberta's nature sites, contact the Fish & Wildlife Div., Dept. of Energy & Natural Resources, Main Floor, North Tower, Petroleum Plaza, 9945 108th St., Edmonton, AB, Canada T5K 2C9 (403/442-2605), For tourism information contact Travel Alberta, Dept. E, PO Box 2500, Edmonton, AB, Canada T5J 2Z4 (from Alberta call 800/222-66501; from U.S. call 403/427-4321).

1. Beaverhall Lake (Ramsar Site). Provincial, 71,781 acres (112 mi²). This site is located near Tofield, about 40 miles southeast of Edmonton. It is an important waterfowl staging area (spring and fall) with more than 200,000 birds regularly using the site each year. During spring migration, more than 150,000 geese stage here, including daily numbers of 50-75,000 Snow Geese and 50-100,000 Greater White-fronted Geese. In fall, 40-70,000 dabbling ducks (mostly Mallards and Northern Pintails) are also present. The lake is also an important waterfowl molting area, with up to 25,000 molting ducks. Sandhill Cranes also stage here in spring migration, with up to 8,000 recorded in late April.

2. Hay-Zama Lakes (Ramsar Site). Provincial, 198,840 acres (226 mi²). This wetland complex attracts large populations of migratory birds during spring and fall migrations. Over 250,000 ducks and 177,000 geese have been observed during a single migration.

It lies on the path of three waterfowl flyways, the Pacific, Central and Mississippi, making it an important molting and staging area for numerous waterfowl species, the primary factor leading to its Ramsar designation.

3. Peace–Athabasca Delta (Ramsar site). Federal, 1,277,746 ac.(1,996 mi²). Located at the western end of Lake Athabasca, its marshes, lakes and mud flats are an important habitat for waterfowl nesting, and provide a staging area for migration. As many as one million ducks, geese and swans pass through this area in the fall.

4. Wood Buffalo National Park; Whooping Crane summer range (Ramsar Site). 4,266 mi²   (16,895 km²). A wetland complex in the boreal forests of northern Alberta and southwestern Northwest Territories, and the only natural nesting habitat for the endangered Whooping Crane. It is owned by the Government of Canada, and is administered by Parks Canada with some input from Indian and Northern Affairs Canada. It is also classified an Important Bird Area and a Ramsar wetland of international importance.

Other protected Alberta sites of possible migratory significance but for which little information was available include Inglewood Migratory Bird Sanctuary (Provincial site, 636 acres), Raven Island National Wildlife Area (Federal site, 240 acres), Richardson Lake Migratory Bird Sanctuary (Provincial site, 50,505 acres), Saskatoon Lake Migratory Bird Sanctuary (Provincial site, 4,534 acres), St. Dennis National Wildlife Area (Federal site, 361 acres), Stalwart National Wildlife Area (Federal site), Tway National Wildlife Area (Federal site), and Webb National Wildlife Area (Federal site).

MANITOBA (MB)

Note: For additional information on Manitoba sites, contact the Manitoba Dept. of Natural Resources, Box 24, 1495 St. James St., Winnipeg, MB, Canada R3H OW9 (204/945-6784). For tourism

information contact Travel Manitoba. Business Development and
Tourism, Winnipeg, MB, Canada R3C 3H8 (204.845-3777, or
800/665-9949).

1. Delta Marsh. This wetland consists of a 7,000-acre open marsh located near the south shore of Lake Manitoba, approximately 15
miles north of the town of Portage la Prairie. The marsh is a wildlife breeding and migration staging area of major importance, especially for Snow and Canada geese, and a variety of ducks. Waterfowl and songbirds are especially abundant in the marsh, either
as breeding residents or seasonal migrants. Warblers migrate
through in waves of thousands. Delta Marsh Bird Observatory, the
most active bird banding station in Canada, bands thousands of
songbirds every year. The marsh was evocatively described in *The*
Canvasback on a Prairie Marsh, a book by H. A. Hochbaum.

2. Oak Hammock Marsh (Ramsar Site). This 5,000-acre marsh is located a few miles north of Winnipeg, Oak Hammock Marsh has
reported 300 species of birds, 193 of these on a regular basis. The
wetland, a restored prairie marsh, is a staging area for waterfowl and shorebirds during migration, when up to 400,000 birds
might be seen in a day. It also hosts a wide variety of songbirds,
including some Manitoba specialties such as Yellow Rail, Nelson's Sparrow, and LeConte's Sparrow. Oak Hammock Marsh
has been designated a Ramsar site in 1987, as a recognized wetland of international importance.

SASKATCHEWAN (SK)

Note: *For additional information on Saskatchewan's nature sites,*
contact the Dept. of Parks, Recreation and Culture, 3211 Albert
Street, Regina, SK, Canada S4S 5W6 (306/787-2700). For tour-
ism information contact Tourism Saskatchewan, 2103 11th Ave.,
Regina, SK, Canada S4P 3V7 (306/5645-2300).

1. Basin & Middle Lake Migratory Bird Sanctuary. Provincial, 34,578
acres (54 mi²). This site near Lake Lenore is important for wa-

terfowl and shorebirds. During surveys completed in 1988 and 1989, an average of 9,578 shorebirds were recorded during three one-day surveys. In addition to shorebirds, over 30,000 ducks have been recorded at this site during the summer molting period (20,000 on Basin Lake, and 10,000 on Middle Lake). See also Lenore Lake account below.

2. Grasslands National Park. 233,1043 ac. Canada's largest are of protected tallgrass prairie, in two blocks. Bird list available (177 spp.). Address: P.O. Box 150, Val Marie, Sask. SON 2TO. Ph. 306/298-2257.

3. Last Mountain Lake National Wildlife Area. Federal/Provincial site. 18,850 acres (29.5 mi²). The north end of Last Mountain Lake is one of the Prairie Provinces' most important waterfowl staging areas, and is a crucial stopover for migratory waterfowl and other water and land birds. During fall up to 50,000 Sandhill Cranes, 450,000 geese, and several hundred thousand ducks may be present. Over 280 species of birds have been recorded here and over 100 of these species have been documented to breed in the area. The area offers important habitat for nine of Canada's 36 rarest birds: the Peregrine Falcon, Piping Plover, Whooping Crane, Burrowing Owl, Ferruginous Hawk, Loggerhead Shrike, Baird's Sparrow, Caspian Tern and Cooper's Hawk. Colonial nesters such as pelicans, cormorants, gulls, terns and grebes are also present. It is a designated Important Bird Area and Ramsar site.

4. Lenore Lake Migratory Bird Sanctuary. Federal/Provincial site. 35,115 acres (56 mi²). Lenore Lake is a partly saline lake near the town of Lake Lenore. It is part of the Lenore Lake Basin, which includes several saline lakes (including Basin & Middle (see above). Lake Lenore is a globally significant site for staging water birds. Tremendous concentrations of birds are present during fall migration, notably 80,000 ducks (mainly Mallards), and 40,000 geese. In the summer about 4,000 ducks (mainly Mallard, Canvasback, and Lesser Scaup) use the lake as a molting area. During periods when good shorebird habitat is available, numbers of shorebirds can be as high as was noted in the spring of 25,000

individuals. Large numbers of Double-crested Cormorants have been documented breeding at the lake, and American White Pelicans also breed at the lake. The nationally endangered Piping Plover has been recorded nesting at the lake in small numbers. It has been designated a provincial Important Bird Area.

5. Old Wives Lake Migratory Bird Sanctuary. Provincial site, 102, 635 acres (160 mi²). Old Wives Lake is a shallow saline lake about 30 km southwest of Moose Jaw. This lake, in conjunction with Reed Lake and Chaplin Lake, forms a site of hemispheric importance in the Western Hemisphere Shorebird Reserve Network. The sanctuary is an important breeding, molting and staging area that attracts large concentrations of ducks, Canada Geese, Snow Geese and Tundra Swans. An estimated 500,000 shorebirds use Old Wives Lake as a stopping point while on migration from their Central or South American wintering grounds to their arctic breeding grounds.

6. Quill Lakes. Provincial site. 252, 527 acres (395 mi²). The Quill Lakes are located immediately north of the town of Wynard in east-central Saskatchewan. During fall migration, the globally threatened Whooping Crane is regularly observed at this site. The Quill Lakes are also significant as a shorebird staging area, with a one day peak count of 197,155 shorebirds being recorded during the spring of 1993. During a 1989-1992 study, several species were recorded in numbers that exceeded one percent of their estimated populations, including Hudsonian Godwit, Least Sandpiper, Baird's Sandpiper, American Avocet and the dowitchers. Especially large numbers of White-rumped Sandpipers and Stilt Sandpipers have been noted. The lakes are also known as an important waterfowl breeding and staging area, with hundreds of thousands of ducks, Sandhill Cranes, Canada Geese, and Snow Geese using the area each fall. The lakes support an exceptional number of breeding Piping Plovers.

7. Redberry Lake Migratory Bird Sanctuary. Provincial site. 25,452 acres (40 mi²). Redberry Lake is a medium-sized saltwater lake near Hafford, and is notable in an area characterized by mostly

freshwater aquatic environments. The lake is an important spring and fall staging ground for ducks and other waterfowl. Its islands serve as the nesting grounds for American White Pelicans, California and Ring-billed gulls, Black and Common terns, and Double-crested Cormorants. White-winged Scoters are also present and presumably breed.

Other protected Saskatchewan sites of possible migratory significance but for which little or no information was available include Duncairn Migratory Bird Sanctuary (Provincial, 6,164 acres), Indian Head Migratory Bird Sanctuary (Federal, 119 acres), Murray Lake Migratory Bird Sanctuary (Provincial, 4,653 acres), Neely Lake Migratory Bird Sanctuary (Provincial, 3,181 acres), Opuntia Lake Migratory Bird Sanctuary (Provincial, 5,567 acres), Scent Grass Lake Migratory Bird Sanctuary (Provincial, 2,505 acres), Upper Rousay Lake Migratory Bird Sanctuary (Provincial, 2,068 acres) and Val Marie Reservoir Migratory Bird Sanctuary (Federal, 2,028 acres).

ARCTIC CANADA

Note: *A permit is required from the Canadian Wildlife Service to visit any sanctuary north of 60º N. latitude (the north boundaries of Alberta and Saskatchewan).*

NORTHWEST TERRITORIES (NT)

Note: *For additional information on Northwest Territories sites, contact the Government of the Northwest Territories, Yellowknife NT, Canada X1A 2L9 (800/661-0788). For birding information contact Travel Arctic, Northwest Territories. Yellowknife, NT, Canada X1A 2L9 (403/873-7200).*

1. Anderson River Delta Migratory Bird Sanctuary. Federal site. 253,000 mi² (655,300 km²). Many species of waterfowl use the

Anderson River, which empties into Amundsen Gulf along Canada's northern coast, for breeding, molting and staging. The western subspecies of Brant breeds on the outer delta. About 2,500 birds, are here from late May through to August or September. The Tundra Swan breeds here in small numbers, and about 1,200 stay to molt in summer. In 1995 at least 3,600 lesser Snow Geese of the Western Central Flyway population bred on islands at the mouth of the Anderson River. Both the Anderson and Mason river deltas support about 1,000 molting Greater White-fronted Geese. Numerous other birds breed, molt, and stage in the area. Large numbers of Long-tailed Ducks. scaup, scoters, dabbling ducks, shorebirds, raptors and songbirds also frequent the delta for breeding or molting.

2. Aulavik National Park. Federal site. 4,750 mi² (12,000 km²), located at north end of Banks Island. Resident birds in this high-arctic park include Common Raven and Rock Ptarmigan. Breeding raptors include Snowy Owl, Gyrfalcon, Rough-legged Hawk and Peregrine Falcon, the latter two of which are migratory.

3. Banks Island Migratory Bird Sanctuaries #1 & #2. Federal sites. Sanctuary #1 = 7,730 mi² (20,200 km²); #2 = 863 mi² (1,430 km²). The largest of Canada's Migratory Bird Sanctuaries, Banks Island is a high-arctic island (above 70º N. lat.) and home to two thirds of the world's population of lesser Snow Geese. For a summary of Banks Island's other birds, see Manning, Höhn, and Macpherson (1956).

4. Cape Parry Important Bird Area, Provincial site. Nationally significant populations of the Common Eider, Glaucous Gull, King Eider, Long-tailed Duck, and Yellow-billed Loon breed on this cape at the northern tip of Perry Peninsula, which projects into Amundsen Gulf. A small Cape Parry Migratory Bird Sanctuary (1.4 mi²), is located here.

5. Tuktut Nogait National Park. Federal site. 16,340 km² (6,470 mi²). Located 100 miles north of the Arctic Circle, along Amundsen Gulf and west of the common boundary of Nunavut and North-

west Territories. This park is a major breeding ground for many arctic-breeding birds. Raptors such as Peregrine Falcon, Rough-legged Hawk, Gyrfalcon and Golden Eagle nest along the steep river canyon walls. Other notable breeding birds include Tundra Swan, Sandhill Crane, Lapland Longspur, Horned Lark, American Golden-plover, and the Yellow-billed, Pacific and Red throated loons.

NUNAVUT (Western) (NU)

Note: This region in Canada's high arctic was named as a territory in 1999, and comprises about a fifth of Canada's total land mass. Its capital is Igauitl, and there are about 26 settlements that total about 25,000 people. Its two largest islands are Baffin Island and Ellesmere Island. Ellesmere Island National Park Preserve is Canada's most northern park.

1. Queen Maud Gulf Migratory Bird Sanctuary. Federal site. 23,846 sq. mi. (61,765 sq. kilometers). This migratory bird sanctuary on mainland Canada's northern coastline is Canada's largest federally protected nature preserve. It is home to one of the planet's greatest concentrations of nesting geese including most of the world population of Ross's geese. The landscape has countless shallow lakes and huge expanses of arctic lowlands. http://www.ibacanada.com/site.jsp?siteID=NU009

2. Polar Bear Pass National Wildlife Area. Federal site. 10.1 sq. mi. (2,620 hectares) On northwestern Bathurst Island (which is located above 75° N. lat,. north of Prince of Wales Island and between Devon and Melville Islands). This wetland ecosystem is a nesting area for 30 species of arctic birds, including King Eider, greater Snow Goose, gulls, jaegers, phalaropes and plovers. It also provides vital habitat for muskoxen, caribou, arctic fox, and marine mammals. It is a Ramsar site, and the second largest wetland area in the Canadian high arctic. http://atlas.nrcan.gc.ca/auth/english/maps/peopleandsociety/nunavut/specialplaces/nationalwildlifeareas/1

3. Seymour Island Migratory Bird Sanctuary. Federal site.18.6 square miles (53 sq. kilometers), including all of Seymour Island (located 30 miles north of Bathurst Island) and the surrounding waters. The high-arctic Seymour Island sanctuary is considered an Important Bird Area. Its bird species include Atlantic Brant, Snowy Owl, and the very rare Ivory Gull. http://www.bsc-eoc. org/iba/site.jsp?siteID=NU045

3

Geographic Distributions and Migration Patterns of Migratory Bird Species in the Central Plains

The following classifications and descriptions of nearly 400 species' distributions and migrations should be regarded as highly tentative, as wintering information on Latin America is often fragmentary, and even for the U.S. detailed migration information is often available only for conspicuous or economically important species. This list excludes all species that breed in the Central Flyway but are essentially non-migratory, those that breed at high-arctic latitudes within the general Central Flyway region as here defined, but migrate via other routes, such as along coastlines or pelagically, and those vagrant species that only very rarely are found within the Central Flyway. Deciding what North American species are essentially "non-migratory" was a subjective process; even such highly cold-tolerant species as the Common Raven (the only bird species that remained throughout the brutal winter of 1804-05 at Lewis & Clark's Fort Mandan) sometimes moves surprising distances, and several species had to be shifted into the "migratory" category after examining banding data. Stotz *et al,* (1996) listed 188 Nearctic species that migrate to the Neotropics but do not breed there, and 141additional Nearctic migrants that also breed there, totaling 329 species. For the Neotropic breeders it is impossible to judge southward migration limits of many Nearctic migrants. Species identified by Stotz *et al.* as Neotropic breeders, or are otherwise known to breed there, are parenthetically indicated in the following summary by the abbreviation "NB". Species whose reported wintering ranges extend into both the Nearctic and Neotropical realms are described as "transitional Neotropic migrants", and those whose historic migration routes (excluding those resulting from recent introductions) included both New World and Old World components are termed Holarctic migrants.

Many species are now known to take over-water "Trans-Gulf" migration routes, flying between the Yucatan Peninsula or the Caribbean region islands and the U.S. Gulf Coast (Lowery, 1945). These are identified as "TG" species. Those birds reported to circumnavigate the Gulf by migrating overland via the east coast of Mexico, or by island-hopping from Yucatan to Florida via Cuba (Gauthreaux, 1971, 1996), are identified as CG species. Probably many eastern-oriented species whose breeding ranges extend west to western Canada (such as the Mourning and Black-and-White warblers) use both CG and TG routes.

The trans-Gulf flight from Yucatan to the northern Gulf Coast is about 450 nautical miles, and for a small passerine bird would require about 15 hours of flying at a ground speed of 30 nautical miles per hour. Such remarkable flights by small passerines often start in early evening, and end by the afternoon of the following day, the birds exploiting tailwinds whenever possible to help conserve energy. By comparison, choosing a probably substantially safer but considerably longer land route around the Gulf would likely take about five to six days to reach the same central Gulf Coast end-point, and thus would require the investment of much more time, and the expenditure of much more energy (Gauthreaux, 1996).

These spring trans-Gulf flights by many North American songbirds originate from a variety of source-points, including the Caribbean islands, the Yucatan Peninsula, and other areas of eastern Mexico located to the south and west of the Bay of Campeche. The northern end-points also range over a broad geographic distance of about 500 miles, extending from the southern tip of coastal Texas to the Florida panhandle. Over many years, the two coastal radar stations that have received the greatest evidence of trans-Gulf migration have been at Houston, Texas, and Lake Charles, Louisiana. Much less is known about the fall return flights, but in the absence of cold fronts the birds evidently move south down the Texas coast and the Florida peninsula. After cold fronts have passed, many birds fly from the upper Gulf Coast to the south Texas and upper Mexican coasts, crossing the northern Gulf during daylight hours (Gauthreaux, 1999).

At least one warbler, the Blackpoll Warbler, evidently flies from the New England coast south directly to northeastern South America via a 2,200-mile non-stop flight (Baird, 1999). It is possible that other warblers, such as the American Redstart, Cape May Warbler and Yellow-rumped Warbler, may also use that route (Nisbet, 1970, Curson, Quinn and Beadle, 1994), and the same may be true of the Connecticut Warbler (Baird, 1999). These phenomenal flights must rank among the most hazardous and strenuous of all passerine migrations and, assuming an average ground speed of 26 mph, might be accomplished in a flying time of 82-88 hours (Williams *et al.*, 1977).

This amazing overseas route represents a flight distance of about 1,500 miles longer than a Florida-based flight would be, but its completion would result in a return to South American wintering areas in the shortest possible time. Similar oceanic flights are typical of many arctic-breeding shorebirds, such as the American Golden-plover, Willet, Whimbrel, Red Knot and Semipalmated Sandpiper. The same migration pattern was used by the ill-fated Eskimo Curlew, whose final remnant population may perhaps have been eliminated by trying to migrate over the West Atlantic Ocean during the frequent autumn storms of the 1930's (Johnsgard, 1980).

Many of the migrating Blackpolls taking this route stop for a time in Bermuda, following a 30-hour flight of about 800 miles from Massachusetts, and at an approximate metabolic cost of about 2.4-2.6 grams of fat. However, the data suggest that those birds bypassing Bermuda have sufficient fat reserves (four grams) to fly the additional 1,350 miles needed to reach South America non-stop (Baird, 1999). Among Nebraska sight records extending from the 1930's to the late 1980's, spring migrant sightings total nearly twelve times as many records as fall sightings (Johnsgard, 2007), suggesting that far fewer Blackpolls migrate south through the Central Flyway during fall than in spring. A similar pattern of differential spring to fall migrant numbers is true in South Dakota (Tallman, Swanson & Palmer, 2002), Kansas (Thompson *et al*, 2012) and Oklahoma (Baumgarten and Baumgarten, 1993).

Indicated national "Population trends" figures refer to estimated percentage rates of annual population increases or decreases for the period 1969–2010 in the U.S. and Canada, based on Breeding Bird Survey data having statistically significant results. "NS" following the percentage figures parenthetically identifies U.S. and Canada population trend data for species lacking statistical significant levels.

The summarized banding information is based on data associated with birds that were banded in Kansas and later recovered or otherwise reported elsewhere, plus those banded elsewhere but recovered or otherwise reported in Kansas (see Table 1). These two data sets were extracted summaries by Thompson *et al,* (2012) and are merged here, to provide more obvious overall patterns of north-south migratory movements within the states and provinces of the Central Flyway. This table excludes data involving inter-state encounters that fall outside of the geographic limits of the Central Flyway as defined here, but does include long-distance banding encounters involving countries beyond the U.S. and Canada.

A long-term tend of gradual warming in the Great Plains, as part of a general global warming trend, has resulted in significantly shorter seasonal migrations for many mid- to high-latitude breeding birds. This trend is illustrated in data from the Audubon's Society's annual Christmas Bird counts in the Great Plains states (North Dakota to Oklahoma plus the Texas panhandle). An analysis of late December bird numbers seen over the four-decade period from 1966 to 2006 for more than 200 migratory species indicates that nearly all of the species have shown a tendency for their early-winter Plains populations to have shifted significantly northward (Johnsgard and Shane, 2009). For many northern species these generally shorter migrations (and longer breeding seasons) might be beneficial, but shifting phenologies that develop among co-evolved species might have undesirable long-term consequences on breeding success..

Many migratory birds exhibit remarkable philopatric (natal- and site-fidelity) tendencies to return to a learned location for breeding, wintering, or staging. To illustrate some notable site-fidelity and longevity records of many highly migratory Central Flyway species, selected band-based longevity records are mentioned in the following species accounts, based on records from the U.S. Bird Banding Laboratory USGS Wildlife Research Center, Laurel, MD. Some examples of inter-state and international movements of birds banded in Kansas (from Thompson *et al.,* 2012) are summarized in Table 1.

Table 1. Banding recoveries & other band encounters in the Central Flyway, from birds banded in Kansas and encountered elsewhere in the Central Flyway or internationally, or banded elsewhere and encountered in Kansas, through 2008 (data from Thompson et al., 2012).

TABLE 1. BANDING RECOVERIES   81

	Mex.	TX	NM	OK	MO	KS	CO	NE	IA	WY	SD	MN	ND	MT	Other*
Snow Goose	-	252	-	20	67	X	-	512	38	-	114	8	79	5	NU, 896, MB 495. ON 114, AL 13 SK 1, Russia 1
Canada Goose	-	57	12	571	702	X	327	1,725	444	28	2,208	943	1,879	19	SK 880, MB 711 AB 74, NU 32 ON 30,
Wood Duck	-	30	-	2	14	X	1	9	65	-	5	68	19	-	SK 3, MB 3, ON 1,
Gadwall	-	37	1	3	-	X	5	9	-	1	12	4	37	11	AB 43. SK 39, MB 9
American Wigeon	-	82	2	32	4	X	3	20	3	2	5	3	37	26	SK 40, NT 14, AB 9, MB 5, YT 1, BC 1,
Am. Black Duck	-	-	-	1	2	X	-	-	-	-	-	2	2	-	ON 4, AB 1, MB 1, SK 1
Mallard	1	890	44	1,709	593	X	857	1,636	382	78	1,361	573	1,396	-574	SK 4.020, AB 2,078 MB 939, NT 346 ON 63, BC 56, AK 16 YT 5
Blue-winged Teal	88	60	1	9	8	X	7	12	25	2	130	31	-188	15	See ***
Northern Shoveler	-	1	-	1	-	X	-	2	1	-	11	9	-	1	SK 24, MB 10, AB 1
Northern Pintail	101	508	6	82	22	X	59	70	20	10	115	48	137	47	See ****
Green-winged Teal	13	131	-	31	8	X	31	18	5	1	24	30	26	5	SK 94, AB 44, MB 36 AK 15, NT 28, ON 9 ON 9
Redhead	-	40	-	13	2	X	7	16	-	6	27	23	24	23	SK 135, MB 71. AB 47 ON 2, BC 2

Table 1, *continued*. Banding recoveries & other band encounters in the Central Flyway, from birds banded in Kansas and encountered elsewhere in the Central Flyway or internationally, or banded elsewhere and encountered in Kansas, through 2008 (data from Thompson et al., 2012).

	Mex.	TX	NM	OK	MO	KS	CO	NE	IA	WY	SD	MN	ND	MT	Other*
Lesser Scaup	1	13	-	5	-	X	1	4	-	-	-	11	1	2	AB 51, AK 18, MB 14, SK 11, NT 5, BC 2, YT 1
Dble-cr. Cormorant	-	-	-	-	-	X	-	-	-	-	10	1	3	1	SK 28, AB 27, MB 2
Am, White Pelican	-	-	-	-	-	X	2	-	-	1	12	24	86	24	MB 6, AB 1,
Great Blue Heron	-	-	-	-	-	X	-	2	-	-	-	1	-	1	SK 21, ON 2,
Great Egret	-	-	-	-	-	X	-	-	-	-	-	1	-	-	
Little Blue Heron	1	2	-	-	-	X	-	1	-	-	-	-	-	-	
Cattle Egret	1	-	-	-	-	X	-	-	-	-	-	-	-	-	
Bk-cr. Night-Heron	-	2	1	-	-	X	1	1	-	-	1	-	1	-	SK 3, Cuba 1,
Turkey Vulture	-	-	-	-	-	X	-	-	-	-	-	-	-	-	SK 1, Ven. 1
Mississippi Kite	-	1	-	-	-	X	-	-	-	-	-	-	-	-	Hond. 1.
Bald Eagle	-	-	-	2	5	X	2	2	-	-	-	2	-	-	ON 3, SK 2
Northern Harrier	-	1	1	2	-	X	-	-	-	-	-	-	-	-	SK 3, Pan. 1
Cooper's Hawk	-	1	-	-	-	X	-	-	-	-	-	1	-	-	SK 1
Swainson's Hawk	-	1	-	1	-	X	4	-	-	-	-	1	-	-	AB 7, Mex. 1, SK 1
Red-tailed Hawk	-	2	-	5	2	X	2	3	1	1	1	1	2	1	AB 21, SK 10, NT 1, YU 1
Ferruginous Hawk	-	-	4	-	-	X	2	2	-	3	2	-	12	2	AB 18, SK 8, MB 2
Golden Eagle	-	-	-	1	-	X	2	6	-	-	1	-	-	-	SK 3, AB 1
American Kestrel	-	-	-	-	-	X	-	-	-	-	-	-	-	-	AB 1, SK 1, YT 1

TABLE 1. BANDING RECOVERIES 83

Table 1, continued. Banding recoveries & other band encounters in the Central Flyway, from birds banded in Kansas and encountered elsewhere in the Central Flyway or internationally, or banded elsewhere and encountered in Kansas, through 2008 (data from Thompson et al., 2012).

	Mex.	TX	NM	OK	MO	KS	CO	NE	IA	WY	SD	MN	ND	MT	Other*
Prairie Falcon	-	-	-	-	-	X	2	-	-	6	-	2	-	-	AB 5, SK 2
American Coot	3	6	-	-	1	X	1	1	-	-	1	-	-	-	MB 2, SK 1
Sandhill Crane	-	1	-	-	-	X	-	-	-	-	-	-	-	-	-
Killdeer	-	1	-	-	-	X	-	-	-	-	-	-	-	-	AB 1
American Avocet	-	-	-	-	-	X	-	-	-	-	-	-	-	-	SA 1
Semipalm. Sandpiper	1	1	-	-	-	X	-	-	-	-	-	-	-	-	So. Am. 26**, C.R. 2, D.R.1, Mart. 1
Western Sandpiper	1	-	-	-	-	X	-	-	-	-	-	-	-	-	BC 4, AK 1
Least Sandpiper	2	1	-	-	-	X	-	-	-	-	-	-	-	-	SK 3
Pectoral Sandpiper	-	1	-	-	-	X	-	-	-	-	-	-	-	—	Mart. 1 , Russia 1
Stilt Sandpiper	-	-	-	-	-	X	-	-	-	-	-	-	-	-	Barb. 2, Arg. 1
Franklin's Gull	-	-	-	-	-	X	-	-	-	-	-	-	2	-	AB 4
Ring-billed Gull	-	-	-	-	-	X	-	-	-	-	-	-	2	-	SA 4, ON 1
Least Tern	-	5	-	-	-	X	-	4	-	-	-	-	-	-	-
Mourning Dove	183	314	9	36	72	X	3	26	16	-	29	7	-13	-	El. Sal. 10, Guat. 9, Nic. 4, Hond. 2, AB 1
Barn Owl	1	2	-	2	-	X	-	-	-	-	-	-	-	-	-
Burrowing Owl	1	-	-	-	-	X	-	-	-	-	-	-	1	-	-
Great Horned Owl	-	-	-	5	1	X	-	-	-	-	-	-	1	-	SK 1
Chimney Swift	-	4	-	3	8	X	-	-	-	-	-	-	-	-	-
R-td Hummingbird	-	-	-	1	-	X	-	-	-	-	-	1	-	-	-
Loggerhead Shrike	-	-	-	-	-	X	-	-	-	-	1	-	-	-	MB 1

Table 1, *continued*. Banding recoveries & other band encounters in the Central Flyway, from birds banded in Kansas and encountered elsewhere in the Central Flyway or internationally, or banded elsewhere and encountered in Kansas, through 2008 (data from Thompson et al., 2012).

	Mex.	TX	NM	OK	MO	KS	CO	NE	IA	WY	SD	MN	ND	MT	Other*
Warbling Vireo	-	-	-	-	-	-	-	-	-	-	-	-	-	-	Guat 1, El. Sal. 1
Blue Jay	-	2	-	5	7	X	1	5	1	-	1	3	-	-	MB 2
Bl.-billed Magpie	-	-	-	-	-	X	1	-	-	-	-	-	-	-	-
American Crow	-	-	-	21	1	X	-	2	1	-	-	1	14	-	SK 18, AB 7, MB 6
Chihuahuan Raven	-	-	-	-	-	X	1	-	-	-	-	-	-	-	-
Purple Martin	-	3	-	6	2	X	-	-	1	-	-	-	-	-	-
Cliff Swallow	-	-	-	-	-	X	-	-	-	-	-	-	-	-	Arg. 1
Barn Swallow	-	-	-	-	-	X	-	-	-	-	-	-	-	-	Guat. 1
Eastern Bluebird	-	1	-	-	-	X	-	-	2	-	-	-	-	-	MB 2
Mount. Bluebird	-	-	-	-	-	X	-	-	-	-	-	-	-	-	AB 1
American Robin	-	4	-	4	7	X	-	4	2	1	5	-	1	-	AK 1
Brown Thrasher	-	5	-	1	-	X	-	1	-	-	-	-	-	-	
Europ. Starling	-	2	1	2	3	X	-	-	9	-	1	4	-	-	
Cedar Waxwing	-	-	-	-	-	X	-	-	-	-	-	2	-	-	
Ovenbird	-	-	-	-	-	X	-	-	-	-	1	-	-	-	
Orange-cr. Warbler	1	-	-	-	-	X	-	-	-	-	-	-	-	-	
Nashville Warbler	-	-	-	-	-	X	-	-	-	-	-	1	-	-	
B.-&-W. Warbler	1	-	-	-	-	X	-	-	-	-	-	-	-	-	
Am. Tree Sparrow	-	-	-	1	-	X	-	2	-	-	1	1	2	-	AB 1, SK 1, MB 1
Vesper Sparrow	-	-	-	-	-	X	-	-	-	-	-	-	-	-	MB 1
Harris's Sparrow	-	2	-	1	-	X	1	2	-	-	6	2	5	-	SK 2, NT 1
Dark-eyed Junco	-	2	-	-	-	X	1	1	1	-	3	2	2	1	MB 1, SK 1

TABLE 1. BANDING RECOVERIES　　　　　　85

Table 1, continued. Banding recoveries & other band encounters in the Central Flyway, from birds banded in Kansas and encountered elsewhere in the Central Flyway or internationally, or banded elsewhere and encountered in Kansas, through 2008 (data from Thompson et al., 2012).

	Mex.	TX	NM	OK	MO	KS	CO	NE	IA	WY	SD	MN	ND	MT	Other*
Red-wngd. Blackbird	-	20	1	8	2	X	-	5	-	-	7	3	9	1	SK 4, AB 3
West. Meadowlark	-	1	-	-	-	X	-	-	-	-	-	-	-	-	-
Yellow-hd. Blackbird	9	3	-	-	-	X	-	1	-	1	9	2	4	-	-
Common Grackle	-	106	-	24	4	X	3	12	1	-	11	2	3	2	ON 1, SK 1
Brown-hd. Cowbird	17	29	-	8	-	X	-	-	1	-	1	3	1	-	SK 1
Purple Finch	-	-	-	1	4	X	-	-	3	-	1	18	-	-	ON 4, MB 2 AB 1, SK 1
House Finch	-	1	-	-	-	X	4	3	4	1	1	2	1	1-	-
Pine Siskin	-	4	-	1	3	X	-	4	-	4	8	8	-	-	AB 1, MB 1 ON 1, SK 1
Am. Goldfinch	-	2	3	5	-	X	-	13	2	2	5	2	2	-	SK 4

Current Standard Canadian abbreviations are: AB, Alberta; MN, Manitoba; NT, Northwest Territory; NU, Nunavut; ON, Ontario; SK, Saskatchewan; YT, Yukon Territory. Other international abbreviations used here are: Arg., Argentina; Barb, Barbados; Col., Colombia; C.R., Costa Rica; D.R., Dominican Republic; El. Sal., El Salvador; Guat., Guatemala; Hisp., Hispaniola; Hond., Honduras; Mart., Martinique; Nic., Nicaragua; Pan., Panama; Ven., Venezuela.

** Includes Guyana/Suriname 16, Venezuela 8, Brazil 1, Ecuador 1, Argentina 1, Alaska 1

*** Includes SK 268, MB 123, AB 121, Col. 56, Cuba 16, Venezuela 16, Guat. 9, MB 7, C.R. 6, Hond. 6, Lesser Antilles 6. Pan. 5, Belize 2, El Sal., 3, Ecuador 2, Hisp. 2, Bahamas 1, Jamaica 1, Nic. 1, NT 1, YT 1.

**** Includes SK 178, MB 120, AB 114, AK 48, NT 46, NU 23, Russia 5, ON 5, BC 4, YT 4, Hond. 1.

Figure 1. Black-bellied Whistling Duck, adults.
All illustrations by author.

Family Anatidae: Swans, Geese & Ducks

Fulvous Whistling-Duck: Southern Nearctic/ Nearctic-Neotropic. Migratory or somewhat nomadic migrant (northern Gulf Coast), or sedentary, ranging south to tropical South America. Widespread and relatively common in the tropical regions of both eastern and western hemispheres. The recent global warming climatic trend might be responsible for the increasing number of sightings of both whistling-duck species in the more northern states of the Central Flyway. Rare in the Central Flyway; U.S. and Canada population trend +1.1% (NS). (NB)

Black-bellied Whistling-Duck: Southern Nearctic/ Nearctic-Neotropic. Migratory (northernmost populations), somewhat migratory, or relatively sedentary, ranging south through Central America to tropical South America. All whistling duck species are tropically distributed, seem to be poorly adapted to temperate-zone breeding (for example, no down is present in whistling-ducks' nests and most species have limited seasonal movements, which are often associated with precipitation patterns rather than temperature). Rare in the Central Flyway. (NB). Figure 1

Greater White-fronted Goose: High-latitude Nearctic/Temperate Nearctic. Long-distance Holarctic migrant, wintering south to northern Mexico. Although this species breeds broadly across the high-arctic tundras of North America and Eurasia, a large portion of the North American population winters in Texas and along the Gulf Coast, migrating in early spring north though the Central Flyway, with perhaps 80 percent of the mid-continent population staging in the central Platte Valley. Birds banded in Alaska, Saskatchewan and northern Canada have been recovered in South Dakota (Tallman, Swanson & Palmer, 2002). Abundant in the Central Flyway; U.S. and Canada population trend +4.7%. Figure 2.

Figure 2. Greater White-fronted Goose, adults and immature.

Snow Goose: High-latitude Nearctic/Temperate Nearctic. Long-distance Nearctic migrant, wintering south to northern Mexico. Evidence from Snow Geese banded in Kansas and reported elsewhere in the Central Flyway states or internationally, or banded

Figure 3. Snow Goose, white and blue morph adults.

elsewhere but reported in Kansas (Table 1) includes the probable importance of Nunavut as a source of the bids migrating through the Central Flyway. Also notable is the occurrence of a banded bird record from as far away as Russia, where about

150,000 birds nest on Belcher Island, off the northeastern coast of Siberia. As of 2012, several million Snow Geese were wintering annually from the lower Missouri-Mississippi Valley west to northeastern Mexico, and migrating through the Central Flyway. Estimates of maximum numbers staging during early spring in the Platte Valley-Rainwater Basin region of Nebraska have ranged as high as seven million birds, with 1-2 million commonly estimated. Roughly 25 percent of the Central Flyway population consists of dark-plumaged (blue-morph) individuals or intermediate (heterozygotic) plumage types (Johnsgard, 2012b). Many birds banded in Alaska and northern Canada have been recovered in South Dakota, and South Dakota-banded birds have been recovered in California and four Mexican states (Tallman, Swanson & Palmer, 2002). Abundant in the Central Flyway. Figure 3.

Ross's Goose: High-latitude Nearctic/Temperate Nearctic. Long-distance Nearctic migrant, wintering south to northern Mexico. This tiny goose is a close relative of the Snow Goose, sometimes hybridizing with it, and comprising about two percent of the combined flocks migrating through Nebraska. If at least two million Snow Geese were present in the Central Flyway in 2012, perhaps Ross's Geese might have then totaled some 40,000 birds. A very few blue-morph Ross's Geese have been reported in the Central Flyway, perhaps the result of hybridization with blue-morph Snow Geese. Some probable wild hybrids between lesser Snow Geese and Ross's geese have been described, and the progressively greater overlap in breeding and wintering ranges of these two close relatives makes much such hybridization likely. Common in the Central Flyway, and apparently increasing rapidly..

Cackling Goose: High-latitude Nearctic/Temperate Nearctic. Long-distance Nearctic migrant, wintering south to northern Mexico. Confusion of this tiny form of white-cheeked goose with the Canada Goose have made surveys and censusing difficult, but available evidence would suggest that 50,000–100,000 of these geese might be present in the Central Flyway. Abundant in the Central Flyway.

Figure 4. Canada Goose, adults.

Canada Goose: Temperate Nearctic/Temperate Nearctic. Variable-distance Nearctic migrant, the northernmost populations the most migratory, and the southernmost ones relatively sedentary. Some migratory populations winter south to northern Mexico.

Evidence from birds banded in Kansas and reported elsewhere in the Central Flyway states or internationally, or banded elsewhere but reported in Kansas (Table 1) includes a substantial movement of birds from Texas north to Nunavut. Many birds banded in northern and central Canada have been recovered in South Dakota, and one banded in South Dakota in 1963 was recovered there in 1997 (Tallman, Swanson & Palmer, 2002). It has been estimated that as of 2009–2011 the North American population of Canada geese was at least 6-7 million birds, of which the Great Plains population probably comprised nearly two million (Johnsgard, 2012a). Abundant and increasing in the Central Flyway, with a national population trend of +9.9%. Figure 4

Trumpeter Swan: Temperate Nearctic/Temperate Nearctic. Variable-distance Nearctic migrant, the northernmost populations the most migratory. Some migratory populations winter south to southern Great Plains; many of these breeding in South Dakota winter in the Nebraska Sandhills, where spring-fed creeks remain fairly ice-free. Increasingly common in the Central Flyway following re-introductions since 1960's. As of 2010, this still-expanding population of re-introduced Trumpeter Swans may have reached 5,000 birds, which breed in scattered northern Plains locations from South Dakota to Ontario (Johnsgard, 2012a).

Tundra Swan: High-latitude Nearctic/Coastal Temperate Nearctic. Long-distance Holarctic migrant, wintering south to eastern and western U.S. coasts. Occasional to rare in the Central Flyway south of South Dakota. Tundra Swans in the Central Flyway migrate mainly from their high-arctic breeding grounds of central Canada south to Montana and North Dakota, from where the flocks separate into two, with a western portion headed southwest through Montana toward wintering areas in California, and the eastern group heading southeast to wintering areas along the mid-Atlantic Coast. Figure 5

Wood Duck: Temperate Nearctic/Southern Nearctic, Long-distance Nearctic migrant, wintering south to northern Mexico. Evidence from birds banded in Kansas and reported elsewhere in the Cen-

Figure 5. Tundra Swan, adults.

tral Flyway states or internationally, or banded elsewhere but reported in Kansas (Table 1) includes that Minnesota and Iowa are probably important sources of birds feeding into the central and southern parts of the Central Flyway. Wood Ducks are much more typical of the Mississippi and Atlantic flyways than the Central Flyway, as they are closely associated with wooded bottomlands and riparian forests. However, with the development of mature woodlands along prairie streams over the past century Wood Ducks have become increasingly common in the Central Flyway region. South Dakota-banded birds have been recovered throughout the Central Flyway, and as far away as Idaho and North Carolina (Tallman, Swanson & Palmer, 2002). Common in the Central Flyway; U.S. and Canada population trend +4.4%. (NB)

Gadwall: Temperate Nearctic/Southern Nearctic. Variable-distance Holarctic migrant, the northernmost populations the most migratory. Some migratory populations winter south to southern Mexico. Evidence from Gadwalls banded in Kansas and reported elsewhere in the Central Flyway states or internationally, or banded elsewhere but reported in Kansas (Table 1) suggests that Alberta, Saskatchewan and North Dakota are probably important sources of Central Flyway birds. Gadwalls are preeminent breeding ducks of prairie wetlands, and thus are among the most common of migratory ducks using the Central Flyway. South Dakota-banded birds have been recovered throughout the Central Flyway, and as far away as California (at 33 years of age) and four Mexican states (Tallman, Swanson & Palmer, 2002). The national estimate of this species' breeding population in 2009 was 3.05 million birds (U. S. Fish & Wildlife Service, 2009). Common in the Central Flyway; U.S. and Canada population trend +2.6%.

American Wigeon: Temperate Nearctic/Southern Nearctic. Variable-distance Nearctic migrant, the northernmost populations the most migratory. Some migratory populations winter south to Central America. Evidence from birds banded in Kansas and reported elsewhere in the Central Flyway states or internationally, or banded elsewhere but reported in Kansas (Table 1) includes the probable importance of the Prairie Provinces and North Dakota as a source for American Wigeons in the Central Flyway. American Wigeons are less clearly associated with prairie wetlands than are Gadwalls, but the two species are often found together during migration. Wigeons are more prone to graze on vegetation along the shoreline than are Gadwalls, and more likely to breed in brushy or somewhat wooded areas. South Dakota-banded birds have been recovered throughout the Central Flyway, and as far away as Alaska and Mexico (Tallman, Swanson & Palmer, 2002). The national estimate of this species' breed-

Figure 6. American Wigeon, adult pair.

ing population in 2009 was 2.47 million birds (U. S. Fish & Wild-life Service, 2009). Common in the Central Flyway; U.S. and Canada population trend –3.3%. Figure 6

American Black Duck: Temperate Nearctic/Temperate Nearctic. Variable-distance Nearctic migrant, the northernmost populations the most migratory. Some migratory populations winter south to southern U.S. Evidence from birds banded in Kansas and reported elsewhere in the Central Flyway states or internationally, or banded elsewhere but reported in Kansas (Table 1) is too limited to show definite geographic patterns. Black Ducks are a species largely associated with East Coast estuaries and woodland-lined marshes, and thus are rare in the Central Flyway. South Dakota band recoveries have come from as far away as Massachusetts, Maryland, and Illinois (Tallman, Swanson & Palmer, 2002). The national estimate of this species' breeding population in 2009 was about 500,000 birds (U. S. Fish & Wildlife Service, 2009). Rare in the Central Flyway; U.S. and Canada population trend –0.4% (NS).

Mallard: Temperate Nearctic/Temperate Nearctic. Variable-distance Holarctic migrant, the northernmost populations the most migratory. Some migratory populations winter south to northern Mexico. Evidence from birds banded in Kansas and reported elsewhere in the Central Flyway states or internationally, or banded elsewhere but reported in Kansas (Table 1) includes a extremely broad diversity in geographic patterns, with very large numbers from Saskatchewan and Alberta, and a surprising number from Alaska. Mallards are the most common ducks of the Central Flyway, indeed perhaps the most common ducks of interior North America and probably also Eurasia. The North American breeding population was recently (U. S. Fish & Wildlife Service, 2009) estimated at 8.5 million birds; it is also probably the most often observed duck species in the Central Flyway, surviving in city ponds and parks as well as in wilderness wetlands. South Dakota-banded birds have been recovered throughout the Central Flyway, and include one banded there in 1952 and recovered in Tennessee in 1983, and one banded in 1996 and recovered in Maryland in 1997 (Tallman, Swanson & Palmer, 2002). Abundant in the Central Flyway; U.S. and Canada population trend +0.2% (NS). (NB [*A. p. diazi*])

Mottled Duck: Southern Nearctic/ Nearctic-Neotropic. Variable-distance transitional Neotropic migrant, the northernmost populations somewhat migratory; the southern ones sedentary. Found mostly along the Gulf Coast, Mottled Ducks are the approximate ecological counterparts of Black Ducks and. like that poorly distinguished species, are very close relatives of Mallards. Like Black Ducks their population has declined as suitable habitats have shrunk and Mallards have largely out-competed them and hybridized with them. An estimate of the Mottled Duck's breeding population in the 1990's was 500-000-800,000 birds (Moorman & Gray, 1994). Uncommon & local in the Central Flyway; U.S. and Canada population trend –3.5%.

Blue-winged Teal: Temperate Nearctic/ Nearctic-Neotropic. Long-distance Nearctic migrant, wintering south from southern U.S. to northern South America. Evidence from birds banded in Kansas and reported elsewhere in the Central Flyway states or internationally, or banded elsewhere but reported in Kansas (Table 1) includes many encounters from as far away as Guyana, Suriname, Venezuela, Brazil and Ecuador. One male, banded in Saskatchewan, was recovered later in Cuba at a minimum age of 23 years, three months. Another male, banded in Manitoba, was later recovered in Mexico at a minimum age of 21 years, 11 months. South Dakota-banded birds have been recovered throughout the Central Flyway, and as far away as Colombia, Venezuela, Peru, Surinam, and Brazil (Tallman, Swanson & Palmer, 2002). Blue-winged Teal are easily the most highly migratory of the breeding ducks of the Central Flyway, leaving early in the fall for warmer latitudes, and arriving back in northern wetlands relatively late in the spring. They are often the most common breeding duck in Great Plains prairie wetlands. The national estimate of this species' breeding population in 2009 was 7.4 million birds (U. S. Fish & Wildlife Service, 2009). Abundant in the Central Flyway; U.S. and Canada population trend –0.1% (NS).

Cinnamon Teal: Temperate Nearctic/ Nearctic-Neotropic. Long-distance Nearctic migrant, the northernmost populations strongly migratory; the southern ones (northern Mexico) sedentary. Winters south to southern Mexico. Like many closely related east-west species pairs (e.g., Indigo/Lazuli buntings, Baltimore/Bull-ock's orioles, Rose-breasted/Black-headed grosbeaks, *etc.*, the Cinnamon Teal has a much shorter migratory route to wintering areas than does its eastern counterpart, the Blue-winged teal. Reasons for these differences are not clear. Cinnamon Teal also seem to have stronger affinities for alkaline wetlands than do Blue-winged Teal, although both may frequently be found nesting on the same wetlands. It is difficult to judge the total population of Cinnamon Teal because of their strong similarities to Blue-winged Teal, but there is one estimate for about the year 2000, was of 260,000 birds (Wetlands International, 2002). Uncommon in the Central Flyway; U.S. and Canada population trend +0.6%. Also resident and breeding in South America. (NB)

Northern Shoveler: Temperate Nearctic/Southern Nearctic. Long-distance Holarctic migrant, the northernmost populations the most migratory. Winters south to northern South America. Evidence from birds banded in Kansas and reported elsewhere in the Central Flyway states or internationally, or banded elsewhere but reported in Kansas (Table 1) includes that Saskatchewan and Manitoba are probably important sources of birds using the Central Flyway. Shovelers forage almost entirely at the water surface, straining out small food items by pumping water out the sides of their enlarged bills, the sides of which have long, closely spaced lamellae that effectively sieve out edible materials. They are common migrants and breeders in the Central Flyway, often associating with blue-winged teal on prairie marshes. South Dakota-banded birds have been recovered throughout the Central Flyway, and as far away as southern Mexico and Cuba (Tallman, Swanson & Palmer, 2002). The national estimate of this species' breeding population in 2009 was 4.38 million birds (U. S. Fish & Wildlife Service, 2009). Common in the Central Flyway; U.S. and Canada population trend +1.4% (NS).

Northern Pintail: Temperate Nearctic/Temperate Nearctic. Variable-distance Holarctic migrant, the northernmost populations the most migratory. Some migratory populations winter south to northern South America. Evidence from Pintails banded in Kan-

Figure 7. Northern Pintail, adult males and female.

sas and reported elsewhere in the Central Flyway states or internationally, or banded elsewhere but reported in Kansas (Table 1) includes the broad-ranging movements of this species, with large numbers of encounters from the Prairie Provinces, but also from Northwest Territories, Yukon Territory, Nunavut, and Russia. One male, banded in Saskatchewan, was recovered there late at a minimum age of 22 years, three months. Northern Pintails, like Mallards, have wide distributions across the northern hemisphere. They are among the most common duck migrants in the Central Flyway, with an estimated North American population of more than three million birds in 2009. Some birds migrate to high-arctic latitudes to breed in tundra habitats, while other breed as far south as grassy wetlands of the southern Great Plains. South Dakota-banded birds have been recovered throughout the Central Flyway, and as far away as Russia, Midway Island, and Panama (Tallman, Swanson & Palmer, 2002). Abundant in the Central Flyway; U.S. and Canada population trend –2.2%. Figure 7.

Green-winged Teal: Temperate Nearctic/Southern Nearctic. Variable-distance Holarctic migrant, the northernmost populations the most migratory. Some migratory populations winter south to southern Mexico. The smallest of North America's surface-feeding ducks the Green-winged Teal is a surprisingly early spring and late fall migrant, often arriving on the central and northern plains with Mallards and Pintails shortly after wetlands become ice-free. Evidence from birds banded in Kansas and reported elsewhere in the Central Flyway states or internationally, or banded elsewhere but reported in Kansas (Table 1) includes a very wide range of geographic encounters, with both Alaska and Mexico showing up frequently. South Dakota-banded birds have been recovered throughout the Central Flyway, and as far away as Alaska and three Mexican states (Tallman, Swanson & Palmer, 2002). The national estimate of this species' breeding population in 2009 was 3.44 million birds (U. S. Fish & Wildlife Service, 2009). Common in the Central Flyway; U.S. and Canada population trend –0.3% (NS).

Canvasback: Temperate Nearctic/Southern Nearctic, Variable-distance Nearctic migrant, the northernmost populations the most migratory. Some migratory populations winter south to central Mexico. Canvasbacks are classic prairie wetland ducks, and are most common as breeders in Canada's Prairie Provinces. There has been a long-term slow population decline as these wetlands have become increasingly rare through drainage and general agricultural pressures. In 2009 the national population was estimated at about 700,000 birds (U. S. Fish & Wildlife Service, 2009). South Dakota-banded birds have been recovered throughout the Central Flyway, and as far away as California, Mexico, and Maryland (Tallman, Swanson & Palmer, 2002). Common in the Central Flyway; U.S. and Canada population trend –0.3% (NS).

Redhead: Temperate Nearctic/Southern Nearctic. Variable-distance Nearctic migrant, the northernmost populations the most migratory. Some migratory populations winter south to northern Central America. Evidence from Redheads banded in Kansas and reported elsewhere in the Central Flyway states or internationally, or banded elsewhere but reported in Kansas (Table 1) indicates that Saskatchewan and Manitoba are important sources of central Flyway birds, One male, banded in Minnesota, was recovered later in Texas at a minimum age of 20 years, seven months. Redheads are much like Canvasback in their behavior and prairie-wetland breeding ecology, but their population has fared somewhat better in recent decades, their national population averaging about one million birds (U. S. Fish & Wildlife Service, 2009). South Dakota-banded birds have been recovered throughout the Central Flyway, and as far away as Alaska, Mexico and several Atlantic Coast states. Some South Dakota birds evidently winter in the Chesapeake region, others in the Gulf Coast, and a few in northern California (Tallman, Swanson & Palmer, 2002). Redheads, more than any other Great Plains breeding duck, seem to fit the conventional Central Flyway pattern the poorest, although large numbers that breed in the Dakotas and southern Canada do winter along the Gulf coast of Texas. Common in the Central Flyway; U.S. and Canada population trend +0.4% (NS). (NB). Figure 8.

Ring-necked Duck: Temperate Nearctic/Southern Nearctic. Variable-distance Nearctic migrant, the northernmost populations the most migratory. Some migratory populations winter south to central Mexico. Ring-necked Ducks often migrate in association

Figure 8. Redhead, adult pair.

with prairie wetland-adapted ducks such as Redheads and Canvasbacks, but breed in more wooded habitats, such as on mus-
keg ponds in Canada's boreal forests. Banded birds have been
recovered in South Dakota from as far away as Maine and Geor-

gia (Tallman, Swanson & Palmer, 2002). The national estimate of this species' breeding population in 2009 was 551,000 (U. S. Fish & Wildlife Service, 2009). Common in the Central Flyway; U.S. and Canada population trend +1.9% (NS).

Greater Scaup: High-latitude Nearctic/Coastal Temperate Nearctic. Variable-distance Holarctic migrant, the northernmost populations the most migratory. Some migratory populations winter south to northern Mexico, largely in coastal habitats. Uncommon in the Central Flyway. This is a relatively rare species in the Central Flyway, as it prefers to use deep waters and coastlines for migration. Banded birds have been recovered in South Dakota from as far away as Maryland and Texas (Tallman, Swanson & Palmer, 2002).

Lesser Scaup: Temperate Nearctic/Nearctic-Neotropic Long-distance Nearctic migrant, wintering south from southern U.S. to southern Central America. Evidence from birds banded in Kansas and reported elsewhere in the Central Flyway states or internationally, or banded elsewhere but reported in Kansas (Table 1) suggests that Alberta and Alaska may be an important sources of Central Flyway birds. Unlike the Greater Scaup, this species is a common migrant in the Central Flyway, and it breeds in the northern parts of the flyway, from the Dakotas north through Alaska's and Canada's boreal forest wetlands. The national estimate of both scaup species' breeding populations in 2009 was 4.2 million birds, of which the Lesser Scaup probably comprised the great majority (U. S. Fish & Wildlife Service, 2009). South Dakota-banded birds have been recovered throughout the Central Flyway, and as far away as Alaska, California and Alabama (Tallman, Swanson & Palmer, 2002). Common in the Central Flyway. (Tallman, Swanson & Palmer, 2002). U.S. and Canada population trend –1.7% (NS).

Surf Scoter: High-latitude Nearctic/Coastal Temperate Nearctic. Long-distance Nearctic migrant, breeding north to Canada's boreal forest and high arctic, wintering mostly along coastlines south to northern Mexico. This is a transitional and rare species in the Central Flyway; most of the birds seen in there are in immature female-like plumage.

White-winged Scoter: High-latitude Nearctic/Coastal Temperate Nearctic. Long-distance Holarctic migrant, breeding north to Canada's boreal forest and high arctic, and wintering mostly along
coastlines south to southern U. S. This is a transitional and rare
species in the Central Flyway; most of the birds seen there are in
female-like plumage.

Long-tailed Duck: High-latitude Nearctic/Coastal Temperate Nearctic. Long-distance Holarctic migrant, breeding in Canada's high
arctic, and wintering south mostly along coastlines to northern
U.S. Rare in the Central Flyway. This is a transitional and fairly
rare species in the Central Flyway; most of the birds seen are in
female-like plumage. Figure 9.

Bufflehead: Temperate Nearctic/Temperate Nearctic. Variable-distance Nearctic migrant, the northernmost Canadian boreal forest
populations the most migratory. Some migratory populations
winter south to central Mexico. One estimate of this species' total breeding population around the year 2000 was roughly one
million birds (Wetlands International, 2002). Birds banded elsewhere have been recovered in South Dakota from as far away
as Saskatchewan and Oklahoma (Tallman, Swanson & Palmer,
2002). Common in the Central Flyway; U.S. and Canada population trend +2.2% (NS).

Common Goldeneye: Temperate Nearctic/Temperate Nearctic. Variable-distance Holarctic migrant, the northernmost populations
of Canada's boreal forest the most migratory. Some migratory populations winter south to northern Mexico. One bird,
banded as a juvenile in Saskatchewan, was recovered later in
the same province at a minimum age of 14 years, three months.
Common in the Central Flyway; U.S. and Canada population
trend +0.9% (NS).

Hooded Merganser: Temperate Nearctic/Temperate Nearctic, Variable-distance Nearctic migrant, the northernmost populations
in Canada's Prairie Provinces the most migratory. Some migratory populations winter south to southern U.S. Uncommon in
the Central Flyway; U.S. and Canada population trend +3.8%.

Figure 9. Long-tailed Duck, adult males and female.

Common Merganser: Temperate Nearctic/Temperate Nearctic. Variable-distance Holarctic migrant, breeding through Canada's boreal forest, the northernmost populations the most migratory.

Some migratory populations winter south to northern Mexico. One estimate of this species' North American breeding population in the early 2000's was about 640,000 birds (Kear, 2005). Common in the Central Flyway; U.S. and Canada population trend −1.0% (NS). (NB)

Red-breasted Merganser: High-latitude Nearctic/Coastal Temperate Nearctic, Variable-distance Holarctic migrant, breeding from Canada's boreal forest north into the high arctic, the northernmost populations along the arctic coast the most migratory. Winters south mostly along coastlines to northern Mexico. This is a transitional species in the Central Flyway; most of the birds seen are in female-like plumage.

Ruddy Duck: Temperate Nearctic/Temperate Nearctic, Variable-distance Nearctic migrant, the northernmost populations in the Northwest Territories the most migratory, and the southernmost probably sedentary. Some migratory populations winter south to southern Mexico. One estimate of this species' total breeding population in the early 2000's was about 500,000 birds (Kear, 2005). Birds banded elsewhere have been recovered in South Dakota from as far away as California and North Carolina (Tallman, Swanson & Palmer, 2002). Common in the Central Flyway; U.S. and Canada population trend +0.1% (NS). Also resident and breeding in South America (O, j. andina and O. j. ferruginea). (NB)

Family Gaviidae: Loons

Red-throated Loon: High-latitude Nearctic/Coastal Temperate Nearctic. Long-distance Holarctic migrant, breeding in the Canadian high arctic, and wintering south to Baja, Mexico. Rare in the Central Flyway; most of the birds seen are in juvenile or winter plumage.

Pacific Loon: High-latitude Nearctic/Coastal Temperate Nearctic. Long-distance Holarctic migrant, breeding in the Canadian high arctic, and wintering south to Baja, Mexico. Rare in the Central Flyway; most of the birds seen are in juvenile or winter plumage.

Common Loon: Temperate Nearctic/Coastal Temperate Nearctic. Long-distance Holarctic migrant, breeding from Minnesota north through Canada's boreal forest, and wintering south to northern Mexico. One bird, banded in Minnesota, was recovered there later at a minimum age of 17 years. Uncommon in the Central Flyway, with all plumage variations seen. U.S. and Canada population trend +0.9% (NS).

Family Podicipedidae: Grebes

Pied-billed Grebe: Temperate Nearctic/Temperate Nearctic. Variable-distance Nearctic migrant, the northernmost populations of north-central Canada are migratory, and the southernmost probably sedentary. Some migratory populations winter south to southern Mexico. Common in the Central Flyway; U.S. and Canada population trend −0.4% (NS). See Figure 10.

Horned Grebe: Temperate Nearctic/Temperate Nearctic. Variable-distance Holarctic migrant, the northernmost populations of north-central Canada most migratory. Some migratory populations winter south to southern U.S. Common in the Central Flyway; U.S. and Canada population trend −2.4%.

Red-necked Grebe: Temperate Nearctic/Temperate Nearctic. Variable-distance Holarctic migrant, the northernmost populations of north-central Canada most migratory. Some migratory populations winter south to southern U.S. Common in the Central Flyway; U.S. and Canada population trend −0.3% (NS). One bird, banded in Minnesota, was recovered there later at a minimum age of 11 years.

Eared Grebe: Temperate Nearctic/Temperate Nearctic. Variable-distance Holarctic migrant, the northernmost populations of central Canada most migratory. Some populations winter south to northern Central America. Common in the Central Flyway; U.S. and Canada population trend −0.2% (NS). (NB)

Western Grebe: Temperate Nearctic/Temperate Nearctic. Variable-distance Nearctic migrant, the northernmost populations of central

Figure 10. American Bittern, Double-crested Cormorant
and Pied-billed Grebe

Canada most migratory, and some southern populations more
sedentary. Migratory populations winter south to southern U.S,
and coastally in Baja Mexico. One bird, banded in Minnesota,
was recovered there later at a minimum age of 11 years. Uncommon in the Central Flyway; U.S. and Canada population trend
(including Clark's Grebe) –0.6% (NS). (NB). Figure 11.

Figure 11. Western Grebe, pair courting.

Clark's Grebe: Temperate Nearctic/Temperate Nearctic. Variable-distance Nearctic migrant, the northernmost populations of southern Canada most migratory, and some southern populations more sedentary. Migratory populations winter south to southern U.S, and coastally in Baja Mexico. Uncommon to rare in the Central Flyway; U.S. and Canada population trend (including Western Grebe) –0.6% (NS). (NB)

Family Phalacrocoracidae: Cormorants

Neotropic Cormorant: Southern Nearctic/Southern Nearctic. Short-distance Nearctic migrant, the northern interior populations in the U.S. probably slightly migratory, and other more southern populations more sedentary. Rare (but apparently increasing rapidly) in the Central Flyway. (NB)

Double-crested Cormorant: Temperate Nearctic/Temperate Nearctic. Variable-distance Nearctic migrant, the northernmost populations of central Canada most migratory, and the southern populations more sedentary. Migratory populations winter south to northern Mexico. Evidence from birds banded in Kansas and reported elsewhere in the Central Flyway states or internationally, or banded elsewhere but reported in Kansas (Table 1) suggests that Alberta and Saskatchewan may be important sources of Central Flyway cormorants. One bird, banded in Saskatchewan, was recovered later in Texas at a minimum age of 17 years, nine months. One bird, banded in South Dakota in 1979, was recovered in South Dakota in 1990, and two others were recovered in Cuba and Florida (Tallman, Swanson & Palmer, 2002). Common in the Central Flyway. Of 93 birds banded elsewhere, the maximum known dispersal distance for an Oklahoma banding documentation was from Alberta (Baumgartner and Baumgartner, 1993). U.S. and Canada population trend +1.1% (NS). (NB). See Figure 10.

Family Anhingidae: Anhingas

Anhinga: Temperate Nearctic/Southern Nearctic. Short-distance Nearctic migrant, the northern interior populations in southern U.S. probably slightly migratory and other populations more sedentary. Rare in the southern Central Flyway; U.S. and Canada population trend +0.8% (NS). (NB)

Figure 12. American White Pelican, adults.

Family Pelecanidae: Pelicans

American White Pelican: Temperate Nearctic/Temperate Nearctic, Variable-distance Nearctic migrant, the northernmost popula- tions of north-central Canada most migratory. Winters south to southern Mexico. Evidence from birds banded in Kansas and re- ported elsewhere in the Central Flyway states or internationally, or banded elsewhere but reported in Kansas (Table 1) suggests that North Dakota, Minnesota and Montana may be important sources of Central Flyway pelicans. Of 78 birds banded elsewhere, the maximum known dispersal distance for an Oklahoma banding documentation was from California (Baumgartner and Baumgartner, 1993), One bird, banded in South Dakota in 1971, was recovered in South Dakota in 1991, and two others were recovered in Mexico (Tallman, Swanson & Palmer, 2002). Common in the Central Flyway; U.S. and Canada population trend +3.0% (NS). (NB), Figure 12.

Family Ardeidae: Bitterns, Egrets & Herons

American Bittern: Temperate Nearctic/Temperate Nearctic. Variable-distance Nearctic migrant, the northernmost populations of central Canada most migratory. Winters south to southern Mexico. Common in the Central Flyway; U.S. and Canada population trend –1.0% (NS). (NB). See Figure 10.

Least Bittern: Temperate Nearctic/Temperate Nearctic. Variable-distance Nearctic migrant, the northernmost populations of southernmost Canada most migratory, and the southern populations more sedentary. Winters south to northern South America. Uncommon in the Central Flyway; U.S. and Canada population trend +0.4% (NS).

Great Blue Heron: Temperate Nearctic/Temperate Nearctic. Variable-distance Nearctic migrant, the northernmost populations of south-central Canada most migratory, and the southern populations more sedentary. Winters south to Mexico. Evidence from birds banded in Kansas and reported elsewhere in the Central Flyway states or internationally, or banded elsewhere but reported in Kansas (Table 1) is too limited to show definite geographic patterns. Of 13 birds banded in Oklahoma or elsewhere, the maximum known dispersal distance for an Oklahoma banding documentation was from Saskatchewan, and the maximum known lifespan of a banded bird was four years (Baumgartner and Baumgartner, 1993). Uncommon in the Central Flyway; U.S. and Canada population trend +0.8%. (NB)

Great Egret: Temperate Nearctic/Temperate Nearctic. Variable-distance Holarctic migrant, the northernmost populations of south-central Canada most migratory, and the southern populations more sedentary. Winters south to Mexico. Evidence from birds banded in Kansas and reported elsewhere in the Central Flyway states or internationally, or banded elsewhere but reported in Kansas (Table 1) is too limited to show definite geographic patterns. Of 13 birds banded in Oklahoma, or banded elsewhere but recovered in Oklahoma, the maximum known dispersal distance for an Oklahoma banding documentation was from Honduras, and the maximum known lifespan of a banded bird was seven years (Baumgartner and Baumgartner, 1993). One bird, banded in Minnesota in 1970, was recovered in South Dakota in 1972 (Tallman, Swanson & Palmer, 2002). Uncommon in the Central Flyway; U.S. and Canada population trend +2.2%. (NB)

Snowy Egret: Temperate Nearctic/Temperate Nearctic. Variable-distance Nearctic migrant, the northernmost populations of the northern Plains states most migratory, and the southern populations more sedentary. Winters south to Mexico. Of 16 birds banded in Oklahoma, or banded elsewhere but recovered in Oklahoma, the maximum known dispersal distance for an Oklahoma banding documentation was from the Pacific Ocean, and the maximum known lifespan of a banded bird was seven years (Baumgartner and Baumgartner, 1993). Uncommon in the Central Flyway; U.S. and Canada population trend –0.2% (NS). (NB)

Little Blue Heron: Southern Nearctic/Southern Nearctic. Variable-distance Nearctic migrant, the northernmost populations or the central Plains states most migratory, and the southern populations more sedentary. Winters south to Mexico. Evidence from birds banded in Kansas and reported elsewhere in the Central Flyway states or internationally, or banded elsewhere but reported in Kansas (Table 1) is too limited to show definite geographic patterns, but records extend south to Mexico. Of 63 birds banded in Oklahoma, or banded elsewhere but recovered in Oklahoma, the maximum known dispersal distance for an Oklahoma banding documentation was from Peru, and the maximum known lifespan of a banded bird was12 years (Baumgartner and Baumgartner, 1993). Uncommon in the Central Flyway; U.S. and Canada population trend –0.9% (NS). (NB)

Tricolored Heron: Southern Nearctic/Southern Nearctic. Variable-distance Nearctic migrant, the northernmost populations of the Gulf Coast slightly migratory, and the southern populations more sedentary. Winters south to Mexico. Rare in the Central Flyway; U.S. and Canada population trend –0.5% (NS). (NB)

Cattle Egret: Temperate Nearctic/Temperate Nearctic. Variable-distance Holarctic migrant, the northernmost populations of the northern Plains states most migratory, and the southern populations more sedentary. Winters south to Mexico. Evidence from birds banded in Kansas and reported elsewhere in the Central Flyway states or internationally, or banded elsewhere but reported in Kansas (Table 1) is too limited to show definite geographic patterns, but records extend south to Mexico. Of 13 birds banded in Oklahoma, the maximum known dispersal distance for an Oklahoma banding documentation was from Mexico, and the maximum known lifespan of a banded bird was eight years (Baumgartner and Baumgartner, 1993). Common in the Central Flyway; U.S. and Canada population trend –0.5% (NS). (NB)

Green Heron: Temperate Nearctic/Temperate Nearctic. Variable-distance Nearctic migrant, the northernmost populations of the northern Plains states most migratory, and the southern populations are relatively sedentary. Winters south to Central America. One bird banded in Oklahoma was recovered in Mexico (Baumgartner and Baumgartner, 1993). Common in the Central Flyway; U.S. and Canada population trend –1.6%.

Black-crowned Night-Heron: Temperate Nearctic/Temperate Nearctic. Variable-distance Nearctic migrant, the northernmost populations of central Canada most migratory, and the southern populations more sedentary. Winters south to Central America. Evidence from birds banded in Kansas and reported elsewhere in the Central Flyway states or internationally, or banded elsewhere but reported in Kansas (Table 1) shows a surprising geographic range of movements, to or from Texas and Cuba in the south, and North Dakota and Saskatchewan in the north. Of 11 birds banded in Oklahoma, or banded elsewhere but recovered in Oklahoma, the maximum known dispersal distance for an Oklahoma banding documentation was from Saskatchewan (Baumgartner and Baumgartner, 1993). Two birds, banded in South Dakota in 1970, were recovered in Guatemala, and one was recovered in Cuba (Tallman, Swanson & Palmer, 2002). Common in the Central Flyway; U.S. and Canada population trend –0.5%. (NB)

Yellow-crowned Night-Heron: Temperate Nearctic/Temperate Nearctic. Variable-distance Nearctic migrant, the northernmost populations of the central Plains states most migratory, and the southern populations more sedentary. Winters south to northern Mexico. Uncommon in the Central Flyway; U.S. and Canada population trend –0.4% (NS). (NB)

Family Threskiornithidae: Ibises & Spoonbills

White Ibis: Southern Nearctic/Southern Nearctic. Short-distance Nearctic migrant, the northern interior populations of the Gulf Coast states slightly migratory, and the southern populations more sedentary. Winters south to northern Mexico. Rare in the Central Flyway; U.S. and Canada population trend +4.8%. (NB)

White-faced Ibis: Temperate Nearctic/Southern Nearctic. Variable-distance migrant, the northernmost populations of the northern Plains states most migratory, and the southern populations more sedentary. Winters south to Central America. Uncommon in the Central Flyway; U.S. and Canada population trend +3.2%. (NB)

Roseate Spoonbill: Southern Nearctic/Southern Nearctic. Short-distance Nearctic migrant, the northern populations of the Gulf Coast states slightly migratory and the southern populations more sedentary. Winters south to northern Mexico. Rare in the Central Flyway; U.S. and Canada population trend +6.1%. (NB)

Family Cathartidae: New World Vultures

Turkey Vulture: Temperate Nearctic/Nearctic-Neotropic. Variable-distance Nearctic migrant, the northernmost populations of southern Canada highly migratory, and the southern populations more sedentary. Some migrants reach South America. Evidence from birds banded in Kansas and reported elsewhere in the Central Flyway states or internationally, or banded elsewhere but reported in Kansas (Table 1) includes a surprising mobility in this vulture, with movements ranging from Saskatchewan to Venezuela. Common in the Central Flyway; U.S. and Canada population trend +4.4%. (NB)

Family Pandionidae: Ospreys

Osprey: Temperate Nearctic/Coastal Southern Nearctic, Variable-distance cosmopolitan migrant, the northernmost populations of north-central Canada highly migratory, and the southern coastal populations more sedentary. Some migrants reach South America. One bird banded in Wisconsin was recovered in Oklahoma (Baumgartner and Baumgartner, 1993). Birds banded elsewhere have been recovered in South Dakota from as far away as Idaho and Washington (Tallman, Swanson & Palmer, 2002). Among six studies of radio-tagged ospreys, migrating birds had average single-day distances covered during autumn that ranged from 88.5 to 180 miles per day (Newton, 2008). Common in the Central Flyway; U.S. and Canada population trend +2.4%. (NB). Figure 13.

Family Accipitridae: Kites, Hawks & Eagles

Swallow-tailed Kite: Southern Nearctic/Central Neotropic. (CG) Long-distance Neotropic migrant, wintering in Central America. Rare and local in the Central Flyway (coastal Texas). U.S. and Canada population trend +5.9%. (NB)

Figure 13. Osprey, adult pair.

White-tailed Kite: Southern Nearctic/Southern Nearctic. Variable-distance Nearctic migrant, the northernmost populations (eastern Texas) slightly migratory and the southern Texas populations more sedentary. Local in the Central Flyway (eastern and the southern Texas). U.S. and Canada population trend –0.2% (NS).

Mississippi Kite. Southern Nearctic/Central Neotropic (CG). Long-distance Neotropic migrant, wintering in South America, south to Argentina and Paraguay. Evidence from birds banded in Kansas and reported elsewhere in the Central Flyway states or internationally, or banded elsewhere but reported in Kansas (Table 1) includes movements as far south as Honduras. One bird, banded in Kansas, was recovered later in Texas at a minimum age of 11 years, two months. One bird banded in Oklahoma was recovered in Guatemala (Baumgartner and Baumgartner, 1993). Autumn counts of migrating raptors in Vera Cruz, Mexico from 2002-2005 produced an average count of 216,000 Mississippi Kites (Newton, 2008), probably representing much of the species' entire population, which is concentrated in the southern Great Plains and southeastern states. Common in the southern Central Flyway (Kansas and south). U.S. and Canada population trend stable at +0.0% (NS).

Bald Eagle: Temperate Nearctic/Temperate Nearctic. Variable-distance Nearctic migrant, the northernmost populations of northern Canada are migratory, and the southern populations are relatively sedentary. Northern birds migrate only far enough to reach open water during winter months. Among three studies of radio-tagged Bald Eagles, migrating birds had average single-day distances covered during autumn that ranged from 7.3 to 91.5 miles per day, while in three spring studies the distances ranged from 50 to 260 miles per day (Newton, 2008). Evidence from birds banded in Kansas and reported elsewhere in the Central Flyway states or internationally, or banded elsewhere but reported in Kansas (Table 1) includes movements to Kansas or from there to as far north as Ontario and Saskatchewan and as far south as Oklahoma. Of four birds banded elsewhere, the maximum known dispersal distance for an Oklahoma banding documentation was from Minnesota (Baumgartner and Baumgartner, 1993). Birds banded elsewhere have been recovered in South Dakota from as far away as Ontario, Minnesota and Texas (Tallman, Swanson & Palmer, 2002). Common in the Central Flyway; U.S. and Canada population trend +5.3%. (NB). Figure 14.

Northern Harrier: Temperate Nearctic/Temperate Nearctic. Variable-distance Nearctic migrant, the northernmost populations of northern Canada are migratory, and the southern populations more sedentary. Winters south to northern Mexico. Evidence from birds banded in Kansas and reported elsewhere in the Cen-

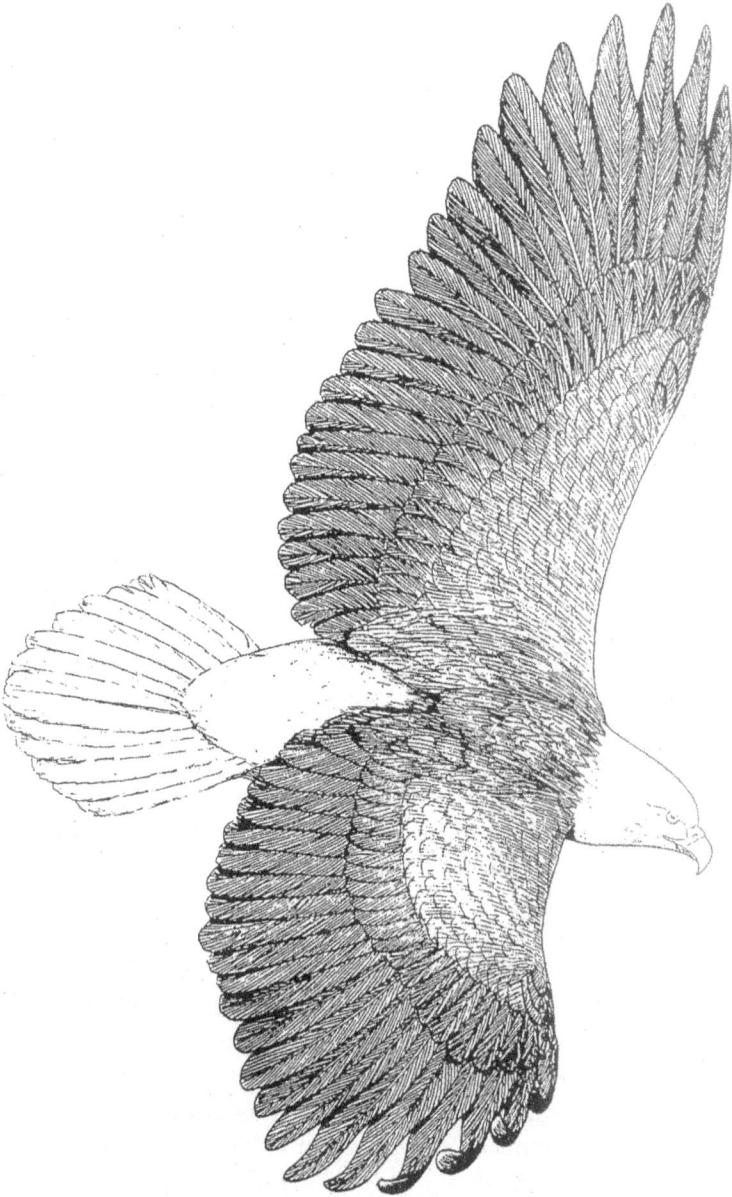

Figure 14. Bald Eagle, adult.

tral Flyway states or internationally, or banded elsewhere but reported in Kansas (Table 1) includes movements to Kansas or from there to as far north as Saskatchewan and as far south as Panama. Of 25 birds banded in Oklahoma, or banded elsewhere but recovered in Oklahoma, the maximum known dispersal distance for an Oklahoma banding documentation was from Alberta (Baumgartner and Baumgartner, 1993). Birds banded in South Dakota have been recovered\from as far away as Texas and Louisiana (Tallman, Swanson & Palmer, 2002). Common in the Central Flyway; U.S. and Canada population trend +0.8%. (NB)

Sharp-shinned Hawk: Temperate Nearctic/Temperate Nearctic. Variable-distance Nearctic migrant, the northernmost populations of northern Canada are migratory, and the southern populations more sedentary. Winters south to Panama. Of two birds banded in Oklahoma, or banded elsewhere but recovered in Oklahoma, the maximum known dispersal distance for an Oklahoma banding documentation was from Minnesota (Baumgartner and Baumgartner, 1993). Common in the Central Flyway; U.S. and Canada population trend +0.9% (NS). (NB)

Cooper's Hawk: Temperate Nearctic/Temperate Nearctic. Variable-distance Nearctic migrant, the northernmost populations of north-central Canada are migratory and the southern populations more sedentary. Winters south to Costa Rica. Evidence from birds banded in Kansas and reported elsewhere in the Central Flyway states or internationally, or banded elsewhere but reported in Kansas (Table 1) includes movements to Kansas or from there to as far north as Saskatchewan and as far south as Texas. Of two birds banded in Oklahoma, or banded elsewhere but recovered in Oklahoma, the maximum known dispersal distance for an Oklahoma banding documentation was from Wisconsin (Baumgartner and Baumgartner, 1993). Another Wisconsin-banded bird was also recovered in South Dakota (Tallman, Swanson & Palmer, 2002). Common in the Central Flyway; U.S. and Canada population trend +2.2%. (NB)

Northern Goshawk: Temperate Nearctic/Temperate Nearctic. Variable-distance Holarctic migrant, the northern populations or

Canada variably migratory or irruptive and the southern populations more sedentary. Newton (2008) reported that, over a period of 120 years, irruptive migrations southward in North America have roughly coincided with those of Great Horned Owls, and are associated with low snowshoe hare numbers. Winters south to Panama. Uncommon in the Central Flyway, wintering or wandering south to Kansas. U.S. and Canada population trend –0.3% (NS).

Common Black-Hawk: Southern Nearctic/Southern Nearctic. Variable-distance Nearctic migrant, the west Texas populations variably migratory probably wintering in northern Mexico; most Mexican populations more sedentary. Rare in the southern Central Flyway, (NB)

Red-shouldered Hawk: Temperate Nearctic/Temperate Nearctic. Variable-distance Nearctic migrant, the northernmost populations of Minnesota and Wisconsin are migratory and the southern populations more sedentary. Winters to central Mexico. One bird banded in Oklahoma was recovered in Texas (Baumgartner and Baumgartner, 1993). Uncommon in the Central Flyway; U.S. and Canada population trend +3.0%. (NB)

Broad-winged Hawk: Temperate Nearctic/Central Neotropic (CG). Long-distance Neotropic migrant, breeding north to central Canada, and wintering in the Central and South America, south to Peru, Bolivia and western Brazil, and east to Venezuela. One Minnesota-banded bird was recovered in South Dakota (Tallman, Swanson & Palmer, 2002). Autumn counts of migrating raptors in Vera Cruz, Mexico from 2002-2005 produced an average count of 2.0 million Broad-winged Hawks (Newton, 2008), probably representing much of the species' entire population, which is concentrated in the deciduous forests of eastern North America. Four radio-tagged adults traveled the approximate 4,200-4,500-mile route from North America to and from their South American wintering grounds in 70 days, averaging about 50 miles per day during both seasons (Haines *et al*, 2003). Uncommon (north) to abundant (south) in the Central Flyway; U.S. and Canada population trend +1.0%. (NB)

Swainson's Hawk: Temperate Nearctic/Southern Neotropic. Long-distance Neotropic migrant, breeding north to south-central Canada, and wintering in South America, from about the Tropic of Capricorn (northern Argentina and the southern Brazil) south to central Argentina. Evidence from birds banded in Kansas and reported elsewhere in the Central Flyway states or internationally, or banded elsewhere but reported in Kansas (Table 1) includes movements to Kansas or from there to as far north as Alberta and Saskatchewan and as far south as Mexico. Of 16 birds banded in Oklahoma, or banded elsewhere but recovered in Oklahoma, the maximum known dispersal distance for an Oklahoma banding documentation was from Brazil (Baumgartner and Baumgartner, 1993). Two Alberta-banded birds and one from Saskatchewan were recovered in South Dakota (Tallman, Swanson & Palmer, 2002). Autumn counts of migrating raptors in Vera Cruz, Mexico from 2002-2005 produced an average count of 1.1 million Swainson's Hawks (Newton, 2008), probably representing much of the species' entire population, which is concentrated in the Great Plains. A group of radio-tagged adults traveled the approximate 8,000-mile route from North America to their South American wintering grounds in 72 days, averaging 115 miles per day. The majority followed a fairly narrow flight-path from eastern Texas, Mexico, Central America, and western South America along the eastern flank of the Andes to Argentine wintering grounds (Fuller, Seegar and Schueck, 1988). Common to abundant in the Central Flyway; U.S. and Canada population trend +0.6%. (NB)

Zone-tailed Hawk: Southern Nearctic/Nearctic-Neotropic. Variable-distance Nearctic migrant, the northernmost west Texas populations are migratory wintering in northern Mexico; Mexican populations probably winter south to Central America. Uncommon in the southern Central Flyway. (NB)

Red-tailed Hawk: Temperate Nearctic/Temperate Nearctic. Variable-distance Nearctic migrant, the northwestern populations breeding in Alaska and northwestern Canada (*B. j. harlani*) the most migratory, wintering south to the southern Great Plains. The proportion of these variably melanistic birds ranges from about

Figure 15. Red-tailed Hawk, adult.

six percent in Montana and North Dakota to about one percent in Nebraska, The semi-albinistic (*krideri*) morph is most common in the central Plains, where they have long been believed to breed, but some of these relatively white-plumaged birds have also been characterized as albinistic *harlani*. The northern and central Great Plains populations are variably migratory, with very large numbers wintering in Oklahoma and northern Texas. Evidence from birds banded in Kansas and reported elsewhere in the Central Flyway states or internationally, or banded elsewhere but reported in Kansas (Table 1) includes movements to Kansas or from there to as far north as Yukon and Northwest Territories and as far south as Texas. One bird, banded in Iowa, was recovered there later at a minimum age of 21 years, six months. Of 11 birds banded in Oklahoma, or banded elsewhere but recovered in Oklahoma, the maximum known dispersal distance for an Oklahoma banding documentation was from Saskatchewan (Baumgartner and Baumgartner, 1993). Birds banded elsewhere have been recovered in South Dakota from as far away as Alberta and Indiana (Tallman, Swanson & Palmer, 2002). Common in the Central Flyway; U.S. and Canada population trend +1.7%. (NB), Figure 15.

Ferruginous Hawk: Temperate Nearctic/Temperate Nearctic. Variable-distance Nearctic migrant, the northernmost populations of southern Canada are migratory, and the southern populations more sedentary. Winters south to northern Mexico. Evidence from birds banded in Kansas and reported elsewhere in the Central Flyway states or internationally, or banded elsewhere but reported in Kansas (Table 1) includes movements to Kansas or from there to as far north as Alberta and Saskatchewan and as far south as Mexico. Of 14 birds banded elsewhere, the maximum known dispersal distance for an Oklahoma banding documentation was from Alberta (Baumgartner and Baumgartner, 1993). Birds banded elsewhere have been recovered in South Dakota from as far away as Utah, and South Dakota-banded birds have been recovered in Texas and two Mexican states (Tallman, Swanson & Palmer, 2002). Uncommon in the Central Flyway; U.S. and Canada population trend +1.4%. Figure 16.

Rough-legged Hawk: High-latitude Nearctic/Temperate Nearctic. Long-distance Holarctic migrant, breeding in the Canadian arctic and wintering in northern and central U.S, south to Oklahoma and northern Texas. Like Snowy Owls, irruptive south-

Figure 16. Ferruginous Hawk, adult.

ward movements in North America average about every 3-5 years, coinciding with population crashes of lemmings (Newton 2008). One bird banded in Ontario was recovered in Oklahoma (Baumgartner and Baumgartner, 1993). Birds banded elsewhere have been recovered in South Dakota from as far away as Colorado and Minnesota (Tallman, Swanson & Palmer, 2002). Common in the Central Flyway.

Golden Eagle: Temperate Nearctic/Temperate Nearctic. Variable-distance Holarctic migrant, the Alaskan and Canadian populations are migratory, and the more southern populations are relatively sedentary. Winters south to Texas and northern Mexico. Evidence from birds banded in Kansas and reported elsewhere

Figure 17. Golden Eagle, adult.

in the Central Flyway states or internationally, or banded elsewhere but reported in Kansas (Table 1) includes movements to Kansas or from there to as far north as Alberta and Saskatchewan and as far south as Oklahoma. One male, banded in Montana, was recovered there later at a minimum age of 28 years, three months. Of four birds banded in Oklahoma, or banded elsewhere but recovered in Oklahoma, the maximum known dispersal distance for an Oklahoma banding documentation was from Alberta (Baumgartner and Baumgartner, 1993). Birds banded elsewhere have been recovered in South Dakota from as far away as Wisconsin (Tallman, Swanson & Palmer, 2002). Uncommon in the Central Flyway; U.S. and Canada population trend +0.1% (NS). Figure 17.

Family Falconidae: Falcons

American Kestrel: Temperate Nearctic/Temperate Nearctic. Variable-distance Nearctic migrant, the Alaskan and Canadian populations are migratory, and more southern U. S. populations are relatively sedentary. Winters south to southern Mexico. Evidence from birds banded in Kansas and reported elsewhere in the Central Flyway states or internationally, or banded elsewhere but reported in Kansas (Table 1) includes movements to Kansas or from there to as far north as Alberta and Saskatchewan and as far south as Texas. Of 15 birds banded in Oklahoma, or banded elsewhere but recovered in Oklahoma, the maximum known dispersal distance for an Oklahoma banding documentation was from Saskatchewan (Baumgartner and Baumgartner, 1993). Birds banded elsewhere have been recovered in South Dakota from as far away as Saskatchewan (Tallman, Swanson & Palmer, 2002). Common in the Central Flyway; U.S. and Canada population trend −1.5%. (NB)

Merlin: Temperate Nearctic/Temperate Nearctic. Variable-distance Holarctic migrant, the Alaskan and Canadian populations are migratory, and the U. S. populations are relatively sedentary. Winters south to northern South America (Venezuela and Ecuador). One bird banded in Manitoba was recovered in Oklahoma (Baumgartner and Baumgartner, 1993). Birds banded elsewhere have been recovered in South Dakota from as far away as Saskatchewan (Tallman, Swanson & Palmer, 2002). Uncommon in the Central Flyway; U.S. and Canada population trend +2.7%.

Gyrfalcon: High-latitude Nearctic/Temperate Nearctic. Variable-distance Holarctic migrant, the high-latitude Canadian populations are migratory, and the western Canada and Alaskan populations are relatively sedentary. Winters south to the northern Plains States, Rare in the Central Flyway. One female, banded in South Dakota, was recovered there later at a minimum age of 13 years, six months.

Peregrine Falcon: Temperate Nearctic/Neotropic. Variable- to long-distance Holarctic migrant, the high-latitude and interior Canadian and U.S. populations highly migratory, and the western coastal populations are relatively sedentary. Winters south to Central and South America (southern Argentina and Chile). One female, banded in Texas, was recovered later in Nicaragua at a minimum age of 19 years, nine months. Of two birds banded elsewhere, the maximum known dispersal distance for an Oklahoma banding documentation was from North Dakota (Baumgartner and Baumgartner, 1993). Among 61 radio-tagged birds, their southward migrations lasted about 50 days and covered about 105 miles per day, while the northern trips averaged 42 days and covered about 120 miles per day. Many of the southward–bound birds moved down along the U.S. Atlantic coast, while many of the northern-bound routes passed through the Great Plains (Fuller *et al.*, 1988). Uncommon in the Central Flyway; U.S. and Canada population trend +1.2% (NS). (NB), Figure 18.

Prairie Falcon: Temperate Nearctic/Temperate Nearctic. Variable-distance Nearctic migrant, the northernmost populations of southern Canada somewhat migratory, and most U. S. populations are relatively sedentary. Winters south to northern Mexico. Evidence from birds banded in Kansas and reported else-

Figure 18. Peregrine Falcon, adult.

where in the Central Flyway states or internationally, or banded elsewhere but reported in Kansas (Table 1) includes movements to Kansas or from there to as far north as Manitoba and Saskatchewan and as far south as Mexico (no numbers provided, but these records were indicated on map by Thompson *et al,* 2012). Of six birds banded in Oklahoma, or banded elsewhere

Figure 19. Prairie Falcon, adult.

but recovered in Oklahoma, the maximum known dispersal distance for an Oklahoma banding documentation was from Alberta (Baumgartner and Baumgartner, 1993). Uncommon in the Central Flyway; U.S. and Canada population trend +1.1% (NS). (NB). Figure 19.

Family Rallidae: Rails, Gallinules & Coots

Yellow Rail: Temperate Nearctic/Temperate Nearctic. Long-distance Nearctic migrant, breeding north to central Canada. Winters south to the Gulf Coast and South Atlantic States coasts. Rare in the Central Flyway.

Black Rail: Temperate Nearctic/Temperate Nearctic. Variable-distance Nearctic migrant, breeding north to central Canada. Those breeding in the central Great Plains probably winter on Texas coast, but very little information on breeding and wintering exists. Rare in the southern Central Flyway.

King Rail: Temperate Nearctic/Temperate Nearctic. Variable-distance Nearctic migrant, the northern populations breeding north to the central Plains states are migratory, and coastal populations are relatively sedentary. Winters south to southern Texas. Uncommon (south) to rare (central) in the Central Flyway; U.S. and Canada population trend –4.0%.

Virginia Rail: Temperate Nearctic/Temperate Nearctic. Variable-distance Nearctic migrant, the populations breeding north to southern Canada are migratory, and the southern populations possibly sedentary. Winters south to southern Mexico. Uncommon in the Central Flyway; U.S. and Canada population trend +0.6% (NS). (NB)

Sora: Temperate Nearctic/Temperate Nearctic. Variable-distance Nearctic migrant, the populations breeding north to southern Canada are migratory, but southernmost populations possibly sedentary. Winters south to Panama. Common in the Central Flyway; U.S. and Canada population trend +0.2% (NS). (NB)

Purple Gallinule: Southern Nearctic/Southern Nearctic. Variable-distance Nearctic migrant, Central Flyway populations of the southern Plains states are migratory. Winters south to Panama. Uncommon in the southern Central Flyway; U.S. and Canada population trend –2.1% (NS). (NB)

Common Gallinule: Temperate Nearctic/Southern Nearctic. Variable-distance Holarctic migrant, the populations of the central and southern Plains states are migratory, but southernmost populations are possibly sedentary. Winters south to Panama. Common in the southern Central Flyway; U.S. and Canada population trend –1.4% (NS). (NB)

American Coot: Temperate Nearctic/Temperate Nearctic. Variable-distance Nearctic migrant, northern populations breeding to northern Canada are migratory, but southern populations more sedentary. Winters south to northern Mexico. Evidence from birds banded in Kansas and reported elsewhere in the Central Flyway states or internationally, or banded elsewhere but reported in Kansas (Table 1) includes movements to Kansas or from there to as far north as Alberta and Saskatchewan and as far south as Mexico. Of 35 birds banded in Oklahoma, or banded elsewhere but recovered in Oklahoma, the maximum known dispersal distance for an Oklahoma banding documentation was from Saskatchewan (Baumgartner and Baumgartner, 1993). Birds banded elsewhere have been recovered in South Dakota from as far away as West Virginia, Michigan and Honduras (Tallman, Swanson & Palmer, 2002). Common in the Central Flyway; U.S. and Canada population trend –0.7% (NS). (NB)

Family Gruidae: Cranes

Sandhill Crane: High-latitude Nearctic/Temperate Nearctic. Variable-distance Nearctic migrant, the Central Flyway population mostly breeding in northeastern Siberia and western Alaska (lesser Sandhills *G. c. canadensis)* and northern Canada (Canadian Sandhills subspecies *G. c. rowani*). These populations are all long-distance migrants, with maximum one-way routes of about 4,000 miles for Russian breeders, but a small percentage of Central Flyway birds (greater Sandhills. *G. c. tabida*) are medium-distance migrants, breeding from Nebraska (rarely) north to southern Canada. Winters south to northern Mexico, primarily in Texas. A radio-tagged adult migrated about 2,000 miles from

Saskatchewan to the Gulf coast of Texas in 103 days, averaging 20 miles per days in fall, but the spring return trip averaged nearly 50 miles per day (Fuller *et al*, 1988). Sandhill and Whooping Cranes typically migrate much faster in spring than during fall, when it is important to reach breeding grounds as early as possible, and when they are not caring for recently fledged young. During their spring stopover in Nebraska's Platte Valley these cranes consume enough food (mostly corn) to build up their fat reserves so that body weight maybe increased by one-third within a month or less. About half of this fat is used during their remaining migration to arctic breeding grounds (Krapu *et al*, 1985). In the fall a similar stopover in the grain fields of southern Canada allows for the buildup of fat sufficient to get them to the Texas wintering grounds. One bird, banded in Nebraska, was recovered later in Texas at a minimum age of 21 years, five months. If this bird nested in western Alaska, where the majority of mid-continent lesser Sandhill Cranes do, it migrated close to 6,000 miles per year, or more than 125,000 miles during its lifetime. If it bred in northeastern Siberia its lifetime cumulative flight distance would likely have approached 170,000 miles. Two others, also banded in Nebraska, were recovered in Texas at minimum ages of 19 years, five months, and 18 years, six months. Of six birds banded in Oklahoma, or banded elsewhere but recovered in Oklahoma, the maximum known dispersal distance for an Oklahoma banding documentation was from Saskatchewan (Baumgartner and Baumgartner, 1993). Abundant in the Central Flyway; U.S. and Canada population trend +4.8%.

Whooping Crane: High-latitude Nearctic/Temperate Nearctic. Long-distance Nearctic migrant, historic population limited to Central Flyway. Winters in coastal Texas. Very rare in the Central Flyway, with about 300 individuals in the Central Flyway population that breeds in Wood Buffalo National Park, Northwest Territories. One bird, banded in the Northwest Territories, was recovered later in Saskatchewan at a minimum age of 28 years, four months. A few color-banded birds are known to have survived more than 30 years and to migrate in extended-family groups that might include up to at least four generations of

Figure 20. Whooping Crane, adult.

closely related individuals (Gill & Johnsgard, 2010). Of 16 birds banded elsewhere, the maximum known dispersal distance for an Oklahoma banding documentation was from Canada's Northwest Territories (Baumgartner and Baumgartner, 1993). At least 55 Canada-banded birds have been recovered in South Dakota, including one almost 12 years old (Tallman, Swanson

& Palmer, 2002). Whooping Cranes in the Wood Buffalo Park–Aransas population migrate much more rapidly than do Sandhill Cranes, averaging only 17 days for the spring migration of more than 2,000 miles, and 30 days for the fall migration. On typical migration days about 245 miles are covered during 7.5 hours of flying time, at an average speed of 32 miles per hour. However, during days with favorable wind 425-490 miles might be covered, at speeds of up to 50 mph (Kuyt, 1992). Except for recently introduced populations, a Central Flyway endemic species. Figure 20.

Family Charadriidae: Plovers

Black-bellied Plover: High-latitude Nearctic/Coastal Neotropic. Long-distance Neotropic migrant, breeding in the high arctic, and wintering mainly in coastal South America, south to central Chile and central Argentina. Common in the Central Flyway.

American Golden-Plover: High-latitude Nearctic/Coastal Southern Neotropic. Long-distance Neotropic migrant, breeding in the high arctic, and wintering in coastal and interior South America, south to southern Brazil, Paraguay and central Argentina. Common in the Central Flyway during spring; the fall migration is performed over water via western Atlantic Ocean.

Snowy Plover: Temperate Nearctic/Coastal Southern Nearctic. Variable-distance Nearctic migrant, interior populations breeding north to the northern Plains states, are migratory, the coastal populations probably more sedentary. Winters south to central Mexico. Uncommon to rare in the Central Flyway. One bird, banded in Kansas, was recovered later in Oklahoma at a minimum age of 11 years. Of three birds banded in Oklahoma, or banded elsewhere but recovered in Oklahoma, the maximum known dispersal distance for an Oklahoma banding documentation was from Kansas (Baumgartner and Baumgartner, 1993).

Wilson's Plover: Coastal Southern Nearctic/Coastal Neotropic. Medium- or short-distance transitional Neotropic migrant, the Gulf

Coast Texas populations are migratory. Winters along Pacific coast, from Mexico to northern Peru. Uncommon in the southern Central Flyway.

Mountain Plover: Temperate Nearctic/Southern Nearctic. Variable-distance Nearctic migrant, interior populations breeding north to Montana are migratory. Winters south from Texas to northern Mexico. Uncommon to rare in the Central Flyway; U.S. and Canada population trend +2.1%.

Semipalmated Plover: High-latitude Nearctic/Coastal Nearctic-Neotropic. Long-distance Neotropic migrant, breeding in the high arctic, and wintering from southern U.S. south to central Chile and central Argentina. One bird, banded in South Dakota in 1990, was recovered in Panama in 1991 (Tallman, Swanson & Palmer, 2002).

Piping Plover: Temperate Nearctic/Coastal Southern Nearctic. Variable-distance Nearctic migrant, the Central Flyway subspecies (*C. m. circumcinctus*) migratory, breeding in the Great Plains states and wintering south along U.S. Gulf Coast, mostly in the area between Aransas and Corpus Christi, Texas. Some color-banded and Nebraska-hatched plovers have also been reported from beaches near New Orleans, Louisiana, Mobile, Alabama, and from Pensacola and Ft. Myers, Florida (Brown & Johnsgard, in prep.). One bird banded in South Dakota was recovered there seven years, ten months later (Tallman, Swanson & Palmer, 2002). Uncommon to rare in the Central Flyway.

Killdeer: Temperate Nearctic/Southern Nearctic. Variable-distance Nearctic migrant, northern populations are migratory, breeding north to northern Canada, and wintering in southern U.S, and northern Mexico, possibly to Central America. Evidence from birds banded in Kansas and reported elsewhere in the Central Flyway states or internationally, or banded elsewhere but reported in Kansas (Table 1) includes movements to Kansas or from there to as far north as Saskatchewan and as far south as Texas. One bird, banded in Kansas, was recovered there later at a minimum age of 10 years, 11 months. Common in the Central Flyway; U.S. and Canada population trend –1.0%. (NB)

Family Recurvirostridae: Stilts & Avocets

Black-necked Stilt: Temperate Nearctic/Southern Nearctic. Variable-distance Nearctic migrant, northern populations breeding north to southern Canada are migratory, wintering in southern U.S, and Mexico, possibly to northern Central America. Gulf Coast populations are relatively sedentary. Uncommon in the Central Flyway; U.S. and Canada population trend +2.4%.

American Avocet: Temperate Nearctic/Southern Nearctic. Variable-distance Nearctic migrant, northern populations breeding north to southern Canada are migratory, wintering in southern U.S, and Mexico, possibly to northern Central America. Gulf Coast populations more sedentary. Evidence from birds banded in Kansas and reported elsewhere in the Central Flyway states or internationally, or banded elsewhere but reported in Kansas (Table 1) includes movements to Kansas or from there to as far north as Saskatchewan. One bird, banded in Kansas, was recovered later in Saskatchewan at a minimum age of nine years. Common in the Central Flyway; U.S. and Canada population trend –0.3% (NS). (NB)

Family Scolopacidae: Sandpipers & Phalaropes

Spotted Sandpiper: Temperate Nearctic/ Nearctic-Neotropic. Long-distance transitional Neotropic migrant, breeding north to northern Canada, and wintering from the southern U.S. south to northern Chile and northern Argentina. Common in the Central Flyway; U.S. and Canada population trend –1.4%.

Solitary Sandpiper: Temperate Nearctic/ Nearctic-Neotropic. Long-distance Neotropic migrant, breeding north to northern Canada, and wintering from the southern U.S. south to Peru and Argentina. Common in the Central Flyway; U.S. and Canada population trend 0.1% (NS).

Greater Yellowlegs: Temperate Nearctic/ Nearctic-Neotropic. Long-distance Neotropic migrant, breeding north to central Canada, and wintering from the southern U.S. south to southern Chile and the southern Argentina. Common in the Central Flyway; U.S. and Canada population trend +2.7% (NS).

Willet: Temperate Nearctic/Neotropic. Long-distance Neotropic migrant, breeding north to south-central Canada, and wintering south to northern Chile and northern Brazil. Common in the Central Flyway during spring; much of the fall migration is performed over water via western Atlantic Ocean, rather than through the Central Flyway; U.S. and Canada population trend –0.4% (NS). (NB). Figure 21.

Figure 21. Willet, adult.

Figure 22. Lesser Yellowlegs, adult.

Lesser Yellowlegs: Temperate Nearctic/Neotropic. Long-distance
    Neotropic migrant, breeding north to northern Canada, and
    wintering south to Chile and Argentina. Common in the Cen-
    tral Flyway; U.S. and Canada population trend –4.5%. Figure 22.

Upland Sandpiper: Temperate Nearctic/Southern Neotropic. Long-
    distance Neotropic migrant, breeding north to northern Canada,
    and wintering south to the grasslands of southern Brazil. Para-
    guay, Uruguay and Argentina. Common in the Central Flyway;
    U.S. and Canada population trend +0.5% (NS).

Whimbrel: High-latitude Nearctic/Coastal Neotropic. Long-distance
    Neotropic migrant, breeding north to northernmost Canada,
    and wintering south to the grasslands of southern Brazil. Par-

aguay, Uruguay and Argentina, with large numbers wintering in southern Patagonia. One bird, banded in Manitoba, was recovered there later at a minimum age of 13 years. Assuming a round-trip migration route from the Canadian arctic to and from southern Patagonia of at least 11,000 miles, this bird flew at least 140,000 miles during its lifetime during its migrations! Common in the Central Flyway during spring; the fall migration is performed over water via the western Atlantic Ocean, rather than through the Central Flyway as is true during spring.

Long-billed Curlew: Temperate Nearctic/Coastal Nearctic-Neotropic. Long-distance Neotropic migrant, breeding north to southern Canada, and wintering south to southern Mexico and Guatemala, occasionally to South America. Common in the Central Flyway; U.S. and Canada population trend +0.4% (NS). Figure 23.

Figure 23. Long-billed Curlew, adult.

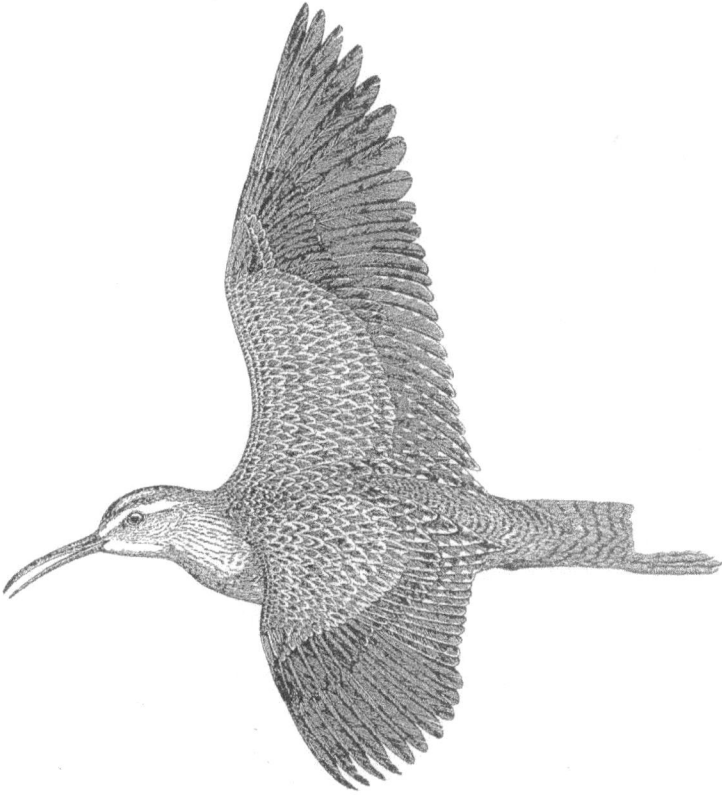

Figure 24. Eskimo Curlew, adult.

Eskimo Curlew. Now probably extinct; previously a High-latitude Nearctic/Coastal Southern Neotropic migrant. It breeding areas in the Canadian arctic were never fully documented, but it wintered in temperate South America, probably in the same areas as the Hudsonian Godwit. Figure 24.

Hudsonian Godwit: High-latitude Nearctic/Coastal Southern Neotropic. Long-distance Neotropic migrant, breeding north to arctic Canada, and wintering south to southern Brazil. Paraguay, Uruguay, Argentina and Chile, with large numbers wintering in southern Patagonia. Common in the Central Flyway during spring; the fall migration is performed over water via western Atlantic Ocean, rather than through the Central Flyway.

Marbled Godwit: Temperate Nearctic/Coastal Northern-Neotropic. Long-distance Neotropic migrant, breeding north to southern Canada, and wintering south to Belize, Costa Rica and Panama, occasionally to northern South America. Common in the Central Flyway; U.S. and Canada population trend –0.3% (NS).

Ruddy Turnstone: High-latitude Nearctic/Neotropic. Long-distance Neotropic migrant, breeding north to arctic Canada, and wintering south throughout coastal Central and South America. Occasional in the Central Flyway. Figure 25.

Figure 25. Ruddy Turnstone, adult.

Red Knot: High-latitude Nearctic/Neotropic. Long-distance Neotropic migrant, breeding north to arctic Canada, and wintering south throughout coastal Central and South America, with large numbers wintering in southern Patagonia. Rare in the Central Flyway during spring; the southward fall migration is performed over water via the western Atlantic Ocean. Many spring migrants heading for arctic Canada follow the Atlantic coast northward, with very few passing up the Central Flyway.

Sanderling: High-latitude Nearctic/Neotropic. Long-distance Neotropic migrant, breeding north to arctic Canada, and wintering south throughout coastal Central and South America, south to Tierra del Fuego. Uncommon in the Central Flyway. Figure 26.

Semipalmated Sandpiper: High-latitude Nearctic/Coastal Neotropic. Long-distance Neotropic migrant, breeding north to arctic Canada, and wintering south to the coasts of Colombia, Venezuela, French Guiana and Brazil. Evidence from birds banded in Kansas and reported elsewhere in the Central Flyway states or internationally, or banded elsewhere but reported in Kansas (Table 1) includes movements to Kansas or from there to as far north as Alaska and as far south as Brazil, Ecuador and Argentina. One bird, banded in Kansas, was recovered there later at a minimum age of 13 years, one month. Common in the Central Flyway during spring; the southward fall migration is over water via the western Atlantic Ocean (Harrington & Morrison, 1979).

Western Sandpiper: High-latitude Nearctic/Coastal Neotropic Long-distance Neotropic migrant, breeding north to arctic Alaska, and wintering south in northern South America, from Guyana west through Venezuela and Colombia, and south through Ecuador to Peru. Evidence from birds banded in Kansas and reported elsewhere in the Central Flyway states or internationally, or banded elsewhere but reported in Kansas (Table 1) includes movements to Kansas or from there to as far north as Alaska and as far south as Mexico. One bird, banded in Kansas, was recovered there later at a minimum age of nine years, two months. Common in the Central Flyway during fall; the spring migration northward closely follows the Pacific Coast (Senner and Martinez, 1982).

Figure 26. Sanderling, adults.

Least Sandpiper: High-latitude Nearctic/Neotropic. Long-distance Neotropic migrant, breeding north to arctic Canada, and wintering south to northern Chile and central Brazil. Evidence from birds banded in Kansas and reported elsewhere in the Central Flyway states or internationally, or banded elsewhere but reported in Kansas (Table 1) includes movements to Kansas or from there to as far north as Saskatchewan and as far south as Mexico. Common in the Central Flyway.

White-rumped Sandpiper: High-latitude Nearctic/Southern Neotropic. Long-distance Neotropic migrant, breeding north to arctic Canada, and wintering south via a fall migration off the Atlantic Coast to eastern South America as far as Tierra del Fuego (Harrington, 1999), with large numbers wintering in southern Patagonia. Cheyenne Bottoms, in western Kansas, is a major spring stopover site. One bird, banded in Kansas, was recovered later in Ontario at a minimum age of six years, two months. Birds banded elsewhere have been recovered in South Dakota from as far away as Manitoba (Tallman, Swanson & Palmer, 2002). Common in the Central Flyway.

Baird's Sandpiper: High-latitude Nearctic/Southern Neotropic. Long-distance Neotropic migrant, breeding north to arctic Canada, and wintering south in southern South America to Tierra del Fuego, mainly coastally but also inland locally in the lower Andes. Jehl (1979) and Harrington (1999) have described the trans-equatorial migrations of this species, which is common during both spring and fall in the Central Flyway. Common in the Central Flyway.

Pectoral Sandpiper: High-latitude Nearctic/Southern Neotropic. Long-distance Neotropic migrant, breeding north to arctic Canada, and wintering south in South America from Peru and the southern Brazil south to southern Argentina and south-central Chile. Evidence from birds banded in Kansas and reported elsewhere in the Central Flyway states or internationally, or banded elsewhere but reported in Kansas (Table 1) includes movements to Kansas or from there to as far north as Russia (northeastern Siberia), and as far south as Martinique. Common in the Central Flyway.

Figure 27. Short-billed Dowitcher, adults.

Dunlin: High-latitude Nearctic/Coastal Southern Nearctic. Long-distance Nearctic migrant, breeding north to arctic Canada, and wintering south coastally to northern Mexico, rarely to Central America. Common in the Central Flyway.

Stilt Sandpiper: High-latitude Nearctic/Southern Neotropic. Long-distance Neotropic migrant, breeding north to arctic Canada, and wintering south in South America from Bolivia and southern Brazil south to northern Argentina and northern Chile.. Evidence from birds banded in Kansas and reported elsewhere in the Central Flyway states or internationally, or banded elsewhere but reported in Kansas (Table 1) includes movements to as far south as Barbados and Argentina. One bird, banded in Kansas, was recovered there later at a minimum age of 11 years, one month. Common in the Central Flyway.

Buff-breasted Sandpiper: High-latitude Nearctic/Southern Neotropic. Long-distance Neotropic migrant, breeding north to arctic Canada, and wintering south to the grasslands of Paraguay, Uruguay and northern Argentina. Common locally in the Central Flyway.

Short-billed Dowitcher: High-latitude Nearctic/Neotropic. Long-distance Neotropic migrant, breeding north to north-central sub-arctic Canada, and wintering (*L. g. hendersoni*) south along both coasts of Central America to northwestern South America, Peru and western Brazil, Common in the Central Flyway. Figure 27.

Long-billed Dowitcher: High-latitude Nearctic/Nearctic-Neotropic. Long-distance Neotropic migrant, breeding north to northwestern arctic Canada and Alaska, and wintering south through the southern U.S. states to Mexico, especially the western Mexican coast and along the Pacific slope. One bird, banded in Kansas, was recovered there later at a minimum age of eight years, four months. Uncommon in the Central Flyway.

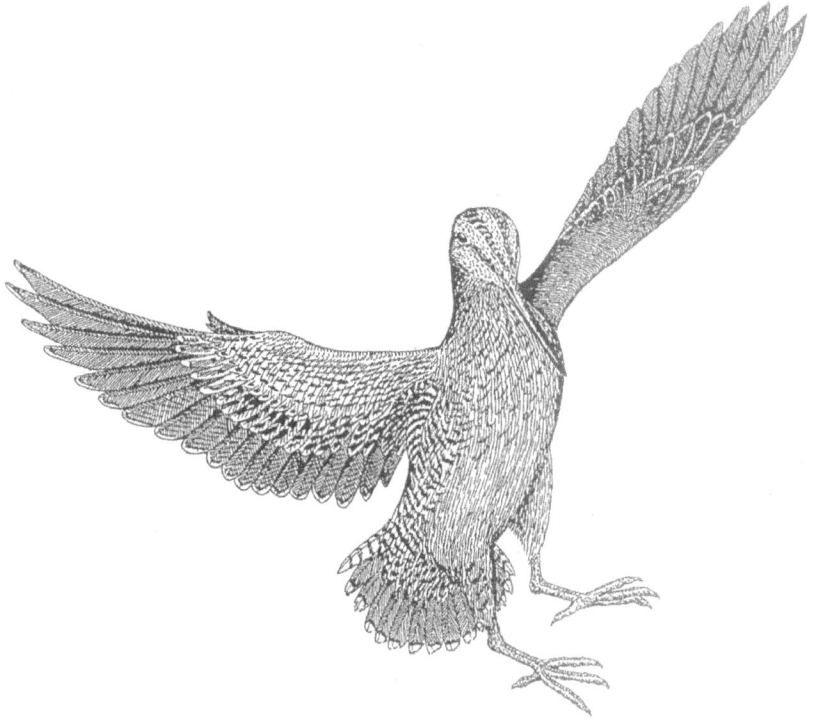

Figure 28. Wilson's Snipe, adult.

Wilson's Snipe: Temperate Nearctic/Nearctic-Neotropic. Long-distance Neotropic migrant, breeding north to arctic Canada, and wintering south through the southern U.S. states to Mexico, occasionally to Central America. Common in the Central Flyway; U.S. and Canada population trend +0.1% (NS). Figure 28.

American Woodcock: Temperate Nearctic/Southern Nearctic Long-distance Neotropic migrant, breeding north to south-central Canada, and wintering south through the southern U.S. states, occasionally to Mexico. Of five birds banded in Oklahoma, or banded elsewhere but recovered in Oklahoma, the maximum known dispersal distance for an Oklahoma banding documentation was from Minnesota (Baumgartner and Baumgartner, 1993). Uncommon in the Central Flyway; U.S. and Canada population trend −1.1% (NS). Figure 29.

Figure 29. American Woodcock, adult.

Wilson's Phalarope: Temperate Nearctic/Southern Neotropic. Long-distance Neotropic migrant, breeding north to arctic Canada, and wintering south to southernmost South America, mostly on high Andean saline lakes and along the Pacific coast. Common in the Central Flyway; U.S. and Canada population trend –0.8% (NS). Figure 30.

Red-necked Phalarope: High-latitude Nearctic/Pelagic Neotropic. Long-distance Neotropic migrant, breeding north to arctic Canada, and wintering pelagically south from Ecuador to southernmost South America, especially off the Pacific coast, where the sea is rich in planktonic foods. Occasional in the Central Flyway.

Red Phalarope: High-latitude Nearctic/Pelagic Neotropic. Long-distance Neotropic migrant, breeding north to arctic Canada, and wintering pelagically south from southern California to southernmost South America, especially off the Pacific coast, where the sea is rich in planktonic foods. Rare in the Central Flyway.

Figure 30. Wilson's Phalarope, adult pair.

Family Laridae: Gulls & Terns

Sabine's Gull: High-latitude Nearctic/Pelagic Neotropic. Long-distance Neotropic migrant, breeding north to high-arctic Canada, and breeding north to arctic Canada, and wintering pelagically south from southern Central America to central Chile, off the Pacific coast. Rare in the Central Flyway.

Bonaparte's Gull: High-latitude Nearctic/Southern Nearctic. Variable-distance Nearctic migrant, breeding north to arctic Canada, and wintering coastally to central Mexico. South Dakota-banded birds have been recovered in North Dakota, Oklahoma and North Carolina (Tallman, Swanson & Palmer, 2002). Common in the Central Flyway.

Laughing Gull. Coastal Temperate Nearctic/Coastal. Short-distance Holarctic migrant, breeding along the Gulf Coast, and with non-breeders reaching central Plains states, migrating south to winter on Gulf Coast. One bird, banded in North Carolina, was recovered in Oklahoma (Baumgartner and Baumgartner, 1993). Rare in the Central Flyway; U.S. and Canada population trend +3.2%. (NB)

Franklin's Gull: Temperate Nearctic/Coastal Southern Neotropic. Long-distance Neotropic migrant, breeding north to north-central Canada, and wintering south to the Pacific coast of South America, south to southern Chile. Evidence from birds banded in Kansas and reported elsewhere in the Central Flyway states or internationally, or banded elsewhere but reported in Kansas (Table 1) includes movements to Kansas or from there to as far north as Alberta. Of nine birds banded in Oklahoma, or banded elsewhere but recovered in Oklahoma, the maximum known dispersal distance for an Oklahoma banding documentation was from Alberta (Baumgartner and Baumgartner, 1993). South Dakota-banded birds have been recovered in Mexico, Central America, Ecuador and Peru (Tallman, Swanson & Palmer, 2002). Common in the Central Flyway; U.S. and Canada population trend −3.9%.

Mew Gull: High-latitude Nearctic/Coastal Temperate Nearctic. Long-distance Holarctic migrant, breeding north to arctic Canada, and wintering south to Baja, Mexico. Rare in the Central Flyway.

Ring-billed Gull: Temperate Nearctic/Southern Nearctic. Variable-distance Nearctic migrant, breeding north to north-central Canada, and wintering south to coast and interior of southern Mexico. Evidence from birds banded in Kansas and reported elsewhere in the Central Flyway states or internationally, or banded elsewhere but reported in Kansas (Table 1) includes movements to Kansas or from there to as far north as Ontario and Saskatchewan. Of 23 birds banded elsewhere, the maximum known dispersal distance for an Oklahoma banding documentation was from Saskatchewan (Baumgartner and Baumgartner, 1993). South Dakota-banded birds have been recovered in the Central Flyway from Saskatchewan and Manitoba to Oklahoma, and in Louisiana (Tallman, Swanson & Palmer, 2002). Common in the Central Flyway; U.S. and Canada population trend +3.0%.

California Gull: Temperate Nearctic/Coastal Nearctic. Variable-distance Nearctic migrant, breeding north to central Canada, and wintering south to coast of northwestern Mexico. One bird, sanded in North Dakota, was recovered in Oklahoma (Baumgartner and Baumgartner, 1993), and another was recovered in South Dakota (Tallman, Swanson & Palmer, 2002). Common in the northern Central Flyway; U.S. and Canada population trend –1.4% (NS).

Herring Gull: Temperate Nearctic/Southern Nearctic. Variable-distance Holarctic migrant, breeding north to arctic Canada. The interior populations are migratory, coastal and the southern populations are sedentary. Winters in southern U.S. and south to Mexico and northern Central America. Of ten birds banded elsewhere, the maximum known dispersal distance for an Oklahoma banding documentation was from Ontario (Baumgartner and Baumgartner, 1993). Birds banded elsewhere have been recovered in South Dakota from as far away as Michigan, Ontario and New Brunswick (Tallman, Swanson & Palmer, 2002). Common in the Central Flyway; U.S. and Canada population trend –2.1%.

Lesser Black-backed Gull: Extralimital/Coastal Nearctic. Long-distance Palaearctic migrant that has recently colonized eastern North America, wintering in coastal and interior U.S. south to Gulf Coast. Rare in the Central Flyway.

Least Tern: High-latitude Nearctic/Neotropic. Long-distance Neotropic migrant, breeding north to the Dakotas, and wintering coastally. Wintering is believed to occur along the Central American coast and along the northern coast of South America, from Venezuela to northeastern Brazil. Evidence from birds banded in Kansas and reported elsewhere in the Central Flyway states or internationally, or banded elsewhere but reported in Kansas (Table 1) includes movements to Kansas or from there to as far north as Nebraska and as far south as Texas. Of 13 birds banded in Oklahoma, or banded elsewhere but recovered in Oklahoma, the maximum known dispersal distance for an Oklahoma banding documentation was from Kansas (Baumgartner and Baumgartner, 1993). Rare in the Central Flyway; U.S. and Canada population trend –2.1% (NS). (NB)

Gull-billed Tern: Temperate Nearctic: Coastal Neotropic. Short-distance Neotropic migrant, breeding on the Gulf Coast, and wintering south to Central America. Local in the southern Central Flyway (Texas coast), where sedentary or nearly so. U.S. and Canada population trend +1.1% (NS).(NB)

Caspian Tern: Temperate Nearctic/Nearctic–Neotropic. Long-distance transitional Neotropic migrant, breeding north to central Canada, and wintering south to Mexico and Central America. One bird, banded in Michigan, was recovered in Oklahoma (Baumgartner and Baumgartner, 1993). Uncommon in the Central Flyway; U.S. and Canada population trend +0.2% (NS). (NB)

Royal Tern: Coastal Temperate Nearctic/Coastal Nearctic. Short-distance Nearctic migrant. Local in the southern Central Flyway (Texas coast), where sedentary or nearly so. (NB)

Sandwich Tern: Coastal Temperate Nearctic/Coastal Nearctic. Short-distance Nearctic migrant. Local in the southern Central Flyway (Texas coast), where sedentary or nearly so. (NB)

Black Tern: Temperate Nearctic/Neotropic. Long-distance Neotropic migrant, breeding north to northern Canada, and wintering on the coast of northern South America east to Suriname, and along western South American coast south to Peru. One bird, banded in Minnesota, was recovered later in El Salvador at a minimum age of eight years, five months. Common in the Central Flyway; U.S. and Canada population trend –2.7% (NS).

Common Tern: Temperate Nearctic/Coastal Neotropic. Long-distance Neotropic migrant, breeding north to north-central Canada, and wintering along the coasts of Central and South America south to Peru and Argentina. Of two birds banded elsewhere, the maximum known dispersal distance for an Oklahoma banding documentation was from Ontario (Baumgartner and Baumgartner, 1993). Birds banded elsewhere have been recovered in South Dakota from as far away as Alberta (Tallman, Swanson & Palmer, 2002). Uncommon in the Central Flyway; U.S. and Canada population trend –0.5% (NS). (NB). Figure 31.

Figure 31. Common Tern, adult.

Forster's Tern: Temperate Nearctic/ Nearctic–Neotropic. Long-distance transitional Neotropic migrant, breeding north to central
Canada, and wintering along the coasts and interior of Mexico and northern Central America. Common in the Central Flyway; U.S. and Canada population trend –0.7% (NS). (NB). Figure 32.

Figure 32. Forster's Tern, adult pair.

Figure 33. White-winged Dove, adult.

Family Columbidae: Pigeons & Doves

White-winged Dove: Southern Nearctic/Southern Nearctic. Short-distance Nearctic migrant in southern Great Plains, breeding north to Oklahoma, and wintering from Texas south. One bird, banded in Texas, was recovered later in El Salvador at a minimum age of 17 years, eight months. Common in the southern Central Flyway; U.S. and Canada population trend +1.1% (NS). (NB). Figure 33.

Mourning Dove: Temperate Nearctic/Southern Nearctic. Variable-distance Nearctic migrant in Great Plains, breeding north to central Canada, and wintering from central Plains states south. Evidence from birds banded in Kansas and reported elsewhere in the Central Flyway states or internationally, or banded elsewhere but reported in Kansas (Table 1) includes movements to Kansas or from there to as far north as Alberta and as far south as Nicaragua. One bird, banded in Colorado, was recovered later in Guatemala at a minimum age of 19 years, four months. Of 979 birds banded in Oklahoma, or banded elsewhere but recovered in Oklahoma, the maximum known dispersal distance for an Oklahoma banding documentation was from Nicaragua (Baumgartner and Baumgartner, 1993). South Dakota-banded birds have been recovered in Maryland, El Salvador and Nicaragua (Tallman, Swanson & Palmer, 2002). Abundant in the Central Flyway; U.S. and Canada population trend +0.4%. (NB)

Family Cuculidae: Cuckoos

Yellow-billed Cuckoo: Temperate Nearctic/Neotropic (TG). Long-distance transitional Neotropic migrant in Great Plains, breeding north to North Dakota, and wintering in South America south to Brazil and Argentina. Common in the Central Flyway; U.S. and Canada population trend –1.6%. (NB)

Black-billed Cuckoo: Temperate Nearctic/Neotropic (TG). Long-distance distance transitional Neotropic migrant in Great Plains, breeding north to southern Canada, and wintering in South America from Venezuela to Peru. Uncommon in the Central Flyway; U.S. and Canada population trend –1.7%.

Family Tytonidae: Barn Owl

Barn Owl: Temperate Nearctic/Southern Nearctic. Variable-distance cosmopolitan migrant. The Great Plains populations (breeding north to Nebraska) probably are mostly short-distance migrants, although Kansas-banded birds have been recovered in Texas and Mexico. Wintering occurs from the south-central Great Plains southward. Evidence from birds banded in Kansas and reported elsewhere in the Central Flyway states or internationally, or banded elsewhere but reported in Kansas (Table 1) includes movements to Kansas or from there to as far north as Ontario and Saskatchewan and as far south as Oklahoma. Of 12 birds banded in Oklahoma, or banded elsewhere but recovered in Oklahoma, the maximum known dispersal distance for an Oklahoma banding documentation was from Louisiana (Baumgartner and Baumgartner, 1993). Birds banded elsewhere have been recovered in South Dakota from as far away as Alberta and Saskatchewan (Tallman, Swanson & Palmer, 2002). Uncommon in the Central Flyway; U.S. and Canada population trend +1.5% (NS). (NB)

Family Strigidae: Typical Owls

Snowy Owl: High-latitude Nearctic/Irruptive. Variable-distance irruptive Holarctic migrant, breeding in arctic Canada, and the largest winter irruptions occurring with arctic prey is scarce, forcing mostly young birds southward. Newton (2008) reported that, over a period of 120 years, such irruptions have occurred on an average of every 3.9 years (Newton, 2008). Birds banded elsewhere have been recovered in South Dakota from as far away as Saskatchewan (Tallman, Swanson & Palmer, 2002). Rare (south) to uncommon (north) in the Central Flyway.

Burrowing Owl: Temperate Nearctic/Southern Nearctic. Variable-distance Nearctic migrant, breeding north to southern Canada, and the northern Great Plains populations the most migratory. Wintering occurs from Texas south to southern Mexico and northern Central America. Evidence from birds banded in Kansas and reported elsewhere in the Central Flyway states or internationally, or banded elsewhere but reported in Kansas (Table 1) includes movement from Mexico to Kansas. South Dakota-banded birds have been recovered from North Dakota to Texas, as well as in Iowa and Arkansas (Tallman, Swanson & Palmer, 2002). One bird, banded in South Dakota, was recovered later in Texas at a minimum age of eight years, eight months. Of three birds banded in Oklahoma, or banded elsewhere but recovered in Oklahoma, the maximum known dispersal distance for an Okla-

Figure 34. Burrowing Owl, adult.

homa banding documentation was from Mexico (Baumgartner
and Baumgartner, 1993). Uncommon in the Central Flyway; U.S.
and Canada population trend –0.8% (NS). (NB). Figure 34.

Great Horned Owl: Temperate Nearctic/Temperate Nearctic. Relative sedentary, but somewhat migratory, and/or irruptive at northern parts of its Nearctic range in northern Canada. Newton (2008) reported that, over a period of 120 years, irruptive migrations southward in North America have roughly coincided with those of Northern Goshawks, and are associated with low snowshoe hare numbers. Evidence from birds banded in Kansas and reported elsewhere in the Central Flyway states or internationally, or banded elsewhere but reported in Kansas (Table 1) includes movements to or from Kansas and as far south as Mexico. Of 25 birds banded in Oklahoma, the maximum known dispersal distance for a recovery was from Kansas (Baumgartner and Baumgartner, 1993). Common in the Central Flyway; U.S. and Canada population trend −0.7%. (NB)

Great Gray Owl: High-latitude Nearctic/Irruptive. Variable-distance irruptive Holarctic migrant, the largest irruptions occurring when boreal-forest prey is scarce, forcing mostly young birds southward. Great Gray Owls, Northern Hawk-owls (*Surnia alula*), Long-eared Owls, Short-eared Owls, and Boreal ("Tengmalm's") Owls all tend to exhibit irruptive southward movements in Europe that average about every 3-5 years, coinciding with cyclic population crashes of microtine prey such as lemmings and voles (Newton 2008). One bird, banded in Alberta, was recovered there later at a minimum age of 12 years, nine months. Very rare in the Central Flyway.

Long-eared Owl: Temperate Nearctic/Temperate Nearctic. Variable-distance Holarctic migrant, the more northerly populations of southern Canada the most migratory; the southern populations are relatively sedentary. Winters south to Texas and northern Mexico. One bird, banded in Saskatchewan, was recovered later in Manitoba at a minimum age of 13 years, seven months. Uncommon in the Central Flyway.

Short-eared Owl: Temperate Nearctic/Temperate Nearctic. Variable-distance Holarctic migrant, the most northerly populations of northern Canada the most migratory; the southern populations are relatively sedentary. One bird, banded in Oklahoma, was recovered in Saskatchewan (Baumgartner and Baumgartner,

1993). Winters south to Texas. Uncommon in the Central Flyway; U.S. and Canada population trend −1.8% (NS). (NB)

Northern Saw-whet Owl: Temperate Nearctic/Temperate Nearctic. Variable-distance Nearctic migrant, the more northerly populations of southern Canada the most migratory; the southern populations are relatively sedentary. Migrations in North America are somewhat irruptive, and are much more marked in some years than others (Newton, 2008). Winters south to the central Great Plains. Uncommon in the Central Flyway. Figure 35.

Figure 35. Saw-whet Owl, adult.

Boreal Owl: High-latitude Nearctic/Irruptive. Variable-distance irruptive Holarctic migrant, breeding north to northern Canada. The largest irruptions occur when microtine rodent (vole) prey is scarce, forcing mostly young birds southward. Very rare in the Central Flyway.

Family Caprimulgidae: Goatsuckers

Lesser Nighthawk: Southern Nearctic/Nearctic-Neotropic. Variable-distance transitional Neotropic migrant, breeding in the Great Plains from southern Texas south, wintering from the southwestern U.S, to northern Bolivia, Paraguay and the southern Brazil. Uncommon in the southern Central Flyway; U.S. and Canada population trend –0.1% (NS). (NB)

Common Nighthawk: Temperate Nearctic/Neotropic (TG). Long-distance Neotropic migrant, breeding from the northern Great Plains south, and wintering in South America south to Argentina. One bird, banded in Ohio, was recovered there later at a minimum age of ten years. Uncommon in the Central Flyway; U.S. and Canada population trend –1.9%. (NB), Figure 36.

Common Poorwill: Temperate Nearctic/ Nearctic-Neotropic. Variable- to short-distance transitional Neotropic migrant, breeding from the northern Great Plains south, and wintering from the southwestern U.S. to central Mexico, Uncommon in the Central Flyway; U.S. and Canada population trend –0.5% (NS).

Chuck-will's-widow: Temperate Nearctic/Neotropic (TG). Variable-distance Neotropic migrant, breeding from the central Great Plains south, and wintering from east-central Mexico to northern Central America. Uncommon in the southern Central Flyway; U.S. and Canada population trend –2.1%.

Eastern Whip-poor-will: Temperate Nearctic/Neotropic (TG). Long-distance Neotropic migrant, breeding from the northern Great Plains south, and wintering from southern Texas to western Panama. Uncommon in the Central Flyway; U.S. and Canada population trend –1.6%. (NB)

Figure 36. Common Nighthawk, adult.

Family Apodidae: Swifts

Chimney Swift: Temperate Nearctic/Neotropic (TG). Long-distance Neotropic migrant, breeding from the northern Great Plains south, and probably wintering in Brazil's western Amazon basin and northern Chile. Evidence from birds banded in Kansas and reported elsewhere in the Central Flyway states or internationally, or banded elsewhere but reported in Kansas (Table 1) includes movements to or from Texas and Oklahoma to Kansas. Of 37 birds banded in Oklahoma or elsewhere, the maximum known dispersal distance for an Oklahoma banding documentation was from Georgia (Baumgartner and Baumgartner, 1993). Common in the Central Flyway. U.S. and Canada population trend –2.2%.

White-throated Swift: Temperate Nearctic/Neotropic. Long-distance transitional Neotropic migrant, breeding from the northern Great Plains south, and wintering from southwestern U.S. to El Salvador and Honduras. Uncommon in the Central Flyway; U.S. and Canada population trend –0.6% (NS).

Family Trochilidae: Hummingbirds

Magnificent Hummingbird: Southern Nearctic/Nearctic–Neotropic. Variable-distance transitional Neotropic migrant, breeding locally in western Texas, and wintering south to western Panama. Rare in the southern Central Flyway.

Blue-throated Hummingbird: Southern Nearctic/Nearctic–Neotropic. Variable-distance transitional Neotropic migrant, breeding locally in western Texas, and wintering in northern Mexico. Rare in the southern Central Flyway.

Lucifer Hummingbird: Southern Nearctic/Nearctic–Neotropic. Variable-distance transitional Neotropic migrant, breeding locally in western Texas, and wintering south to central Mexico. Rare in the southern Central Flyway.

Ruby-throated Hummingbird: Temperate Nearctic/Nearctic–Neotropic (TG). Variable-distance transitional Neotropic migrant, breeding from the northern Great Plains south, and wintering south to Panama. Evidence from birds banded in Kansas and reported elsewhere in the Central Flyway states or internationally, or banded elsewhere but reported in Kansas (Table 1) includes movements to or from Oklahoma to Kansas, and from Kansas to Minnesota. One bird, banded in Oklahoma, was recovered there later at a minimum age of nine years, and another banded in Oklahoma was recovered in Texas (Baumgartner and Baumgartner, 1993). Common in the Central Flyway; U.S. and Canada population trend +1.9%. Figure 37.

Black-chinned Hummingbird: Temperate Nearctic/Nearctic–Neotropic. Variable -distance transitional Neotropic migrant, breeding from the southern Great Plains south, and wintering south

Figure 37. Ruby-throated Hummingbird, female at nest.

to western and central Mexico. One bird, banded in Texas, was recovered there later at a minimum age of 11 years, two months. Uncommon in the southern Central Flyway; U.S. and Canada population trend +1.2%. (NB)

Calliope Hummingbird. Extralimital/Nearctic–Neotropic. Variable-distance transitional Neotropic migrant, breeding in the Rocky Mountains and intermountain West, migrating through the western edge of the Central Flyway, and wintering south to western and central Mexico. One bird, banded in Montana, was recovered there later at a minimum age of six years, 11 months. Uncommon in the southern Central Flyway; U.S. and Canada population trend +0.1% (NS). (NB)

Broad-tailed Hummingbird: Extralimital/Nearctic–Neotropic. Variable-distance transitional Neotropic migrant, breeding in the Rocky Mountains and intermountain West, and wintering south to western and central Mexico. One bird, banded in Colorado, was recovered there later at a minimum age of 12 years, two months. Uncommon in the southern Central Flyway; U.S. and Canada population trend –1.1% (NS).

Rufous Hummingbird: Extralimital/Nearctic–Neotropic. Variable-distance transitional Neotropic migrant, breeding in the intermountain West and Pacific Northwest, and wintering south to western and central Mexico, and east rarely to the Atlantic coast. Many Rufous Hummingbirds migrate south along the front range of the Rockies, at the western edge of the Central Flyway, and some regularly cross the Central Flyway to winter along the Gulf Coast, from Alabama west to Mexico (Calder, 1999). Uncommon in the southern Central Flyway; U.S. and Canada population trend +2.2%.

Family Alcedinidae: Kingfishers

Belted Kingfisher: Temperate Nearctic/Nearctic-Neotropic (TG). Variable-distance transitional Neotropic migrant, breeding north to northern Canada and Alaska, wintering as far south as needed to find open water. Neotropic populations winter to northern South America. Common in the Central Flyway; U.S. and Canada population trend –1.4%.

Family Picidae: Woodpeckers

Lewis's Woodpecker: Temperate Nearctic/Temperate Nearctic. Variable-distance Nearctic migrant, breeding from the western edge of the Great Plains south, the more northerly populations the most migratory; the southern populations are relatively sedentary. Winters south to New Mexico and Arizona. Uncommon in the Central Flyway; U.S. and Canada population trend –1.8%.

Red-headed Woodpecker: Temperate Nearctic/Temperate Nearctic. Variable-distance Nearctic migrant, breeding from the northern Great Plains south, the more northerly populations the most migratory; the southern populations are relatively sedentary. Winters south to southern Texas. Birds banded elsewhere have been recovered in South Dakota from as far away as Illinois, and a South Dakota-banded bird has been recovered in Illinois (Tallman, Swanson & Palmer, 2002). Common in the Central Flyway. U.S. and Canada population trend +2.7%.

Yellow-bellied Sapsucker: Temperate Nearctic/ Nearctic-Neotropic. Variable-distance transitional Neotropic migrant, breeding from the northern Great Plains south; all U.S. populations apparently variously migratory. Winters from the central Great Plains south through the Neotropics to Panama, also in the Caribbean region. The Neotropics-wintering birds reportedly are trans-Gulf migrants. Females winter farther south than do males, and return to breeding areas later. One bird, banded in Oklahoma, was recovered in Minnesota (Baumgartner and Baumgartner, 1993). Common in the Central Flyway; U.S. and Canada population trend +0.8% (NS). (NB)

Red-naped Sapsucker: Temperate Nearctic/Temperate Nearctic. Variable-distance Nearctic migrant, breeding from the westernmost Great Plains south; all populations variously migratory. Winters from the southwestern states south to western Mexico. Uncommon to rare in the western Central Flyway; U.S. and Canada population trend +1.5%. One bird, banded in Wyoming, was recovered there later at a minimum age of four years, nine months.

Williamson's Sapsucker: Temperate Nearctic/Temperate Nearctic. Variable-distance Nearctic migrant, breeding from the westernmost Great Plains south; all populations variously migratory. Winters from the southwestern states south to western Mexico.

Uncommon to rare in the western Central Flyway; U.S. and Canada population trend –0.1% (NS).

Northern Flicker: Temperate Nearctic/Temperate Nearctic. Variable-distance Nearctic migrant, breeding from the northern Great Plains south, the more northerly populations the most migratory; the southern populations are relatively sedentary. Winters from the southern Great Plains states south to northern Mexico. Neotropic populations breed and winter south to Nicaragua. Of 25 birds banded in Oklahoma, or banded elsewhere but recovered in Oklahoma, the maximum known dispersal distance for an Oklahoma banding documentation was from Saskatchewan (Baumgartner and Baumgartner, 1993). South Dakota-banded birds have been recovered in Texas (Tallman, Swanson & Palmer, 2002). Common in the Central Flyway; U.S. and Canada population trend –1.4% (NS).

Family Tyrannidae: American Flycatchers

Olive-sided Flycatcher: Temperate Nearctic/Central Neotropic (TG). Long-distance Neotropic migrant, breeding north to northern Alaska and northern Canada, wintering from southern Mexico to northwestern South America, from Suriname south to Bolivia. Uncommon in the Central Flyway; U.S. and Canada population trend +3.4%.

Western Wood-Pewee: Temperate Nearctic/Central Neotropic. Long-distance Neotropic migrant, breeding north to central Alaska and northwestern Canada, wintering in northwestern South America from Venezuela south to Bolivia. Uncommon in the western Central Flyway; U.S. and Canada population trend –1.6%. (NB)

Eastern Wood-Pewee: Temperate Nearctic/Central Neotropic (TG). Long-distance Neotropic migrant, breeding north to southern Canada, wintering in northwestern South America from Venezuela south to Bolivia. Common in the Central Flyway; U.S. and Canada population trend +1.3%.

Yellow-bellied Flycatcher: Temperate Nearctic/Nearctic-Neotropic Long-distance Neotropic migrant, breeding north to central

Alaska and northwestern Canada, wintering from eastern Mexico to Panama. Uncommon in the Central Flyway; U.S. and Canada population trend +1.9%.

Acadian Flycatcher: Temperate Nearctic/Central Neotropic. Long-distance Neotropic migrant, breeding north to the Great Lakes states, wintering from Nicaragua south to Ecuador. Uncommon in the eastern Central Flyway; U.S. and Canada population trend –0.4% (NS).

Alder Flycatcher: Temperate Nearctic/Central Neotropic. Long-distance Neotropic migrant, breeding north to central Alaska and northwestern Canada, wintering in western South America south possibly to northern Argentina. One bird, banded in Alberta, was recovered there later at a minimum age of seven years, one month. Of two birded banded in western Nebraska, one was recaptured the same year near Tok, Alaska, and another at Fairbanks (Scharf et al, 2008). Uncommon in the Central Flyway; U.S. and Canada population trend (including Willow Flycatcher) –1.3%.

Willow Flycatcher: Temperate Nearctic/Central Neotropic. Long-distance Neotropic migrant, breeding north to southern Canada, wintering in western South America, south possibly to northern Argentina. Common in the Central Flyway; U.S. and Canada population trend (including Alder Flycatcher) –1.3%.

Least Flycatcher: Temperate Nearctic/Nearctic-Neotropic (CG). Long-distance Neotropic migrant, breeding north to northwestern Canada, wintering from Mexico to Costa Rica. Common in the Central Flyway; U.S. and Canada population trend –1.7%.

Black Phoebe: Southern Nearctic/Neotropic. Variable-distance Neotropic migrant, breeding north to northern Texas, wintering from the southwestern states to southern Mexico. Common in southernmost Central Flyway.

Eastern Phoebe: Temperate Nearctic/Nearctic–Neotropic (TG). Variable-distance Neotropic migrant, breeding north to northwestern Canada, wintering from the southwestern states to central Mexico. Common in the Central Flyway; U.S. and Canada population trend +0.7% (NS).

Say's Phoebe: Temperate Nearctic/Nearctic–Neotropic. Variable-distance Neotropic migrant, breeding north to northern Alaska and northwestern Canada, wintering from the southwestern states to southern Mexico. Common in the western Central Flyway; U.S. and Canada population trend +0.4% (NS). (NB)

Vermilion Flycatcher: Southern Nearctic/Neotropic. Variable-distance Neotropic migrant, breeding north to central Texas, wintering from the southwestern states to Nicaragua. Common in the southern Central Flyway; U.S. and Canada population trend +0.1% (NS).

Ash-throated Flycatcher: Temperate Nearctic/Nearctic–Neotropic. Variable-distance Neotropic migrant, breeding north to Oregon, wintering from the southwestern states to Nicaragua. Common in southwestern Central Flyway; U.S. and Canada population trend +1.1%. (NB)

Great Crested Flycatcher: Temperate Nearctic/Nearctic-Neotropic (TG). Long-distance Neotropic migrant, breeding north to northern Saskatchewan, wintering from southern Mexico to northwestern South America, from Venezuela to Colombia. Common in the Central Flyway; U.S. and Canada population trend –0.1% (NS).

Brown-crested Flycatcher: Southern Nearctic/Neotropic. Variable-distance Neotropic migrant, breeding north to southern Texas, wintering from southern Mexico to Costa Rica. Local in the southern Central Flyway.

Tropical Kingbird: Southern Nearctic/Neotropic. Short-distance Neotropic migrant (or possible permanent resident), breeding north to southernmost Texas, wintering or permanent resident from northern Mexico to Peru. Local in extreme southern Central Flyway.

Couch's Kingbird: Southern Nearctic/Neotropic. Short-distance Neotropic migrant, breeding north to southern Texas, wintering from northernmost Mexico to southern Mexico. Local in extreme southern Central Flyway; U.S. and Canada population trend +8.6%.

Cassin's Kingbird: Temperate Nearctic/Nearctic–Neotropic. Variable-distance transitional Neotropic migrant, breeding north to

Montana; wintering from northwestern Mexico to northern Central America. Common in the western Central Flyway; U.S. and Canada population trend +0.2% (NS).

Western Kingbird: Temperate Nearctic/Nearctic–Neotropic (TG). Long-distance transitional Neotropic migrant, breeding north to southern Canada, and wintering from northern Mexico to Nicaragua. One bird, banded in South Dakota, was recovered there later at a minimum age of six years, 11 months. Common in the Central Flyway; U.S. and Canada population trend +0.5%. (NB)

Eastern Kingbird: Temperate Nearctic/Neotropic (TG). Long-distance transitional Neotropic migrant, breeding north to northwestern Canada, and wintering in South America from Venezuela and Ecuador south to southern Brazil and Argentina. Common in the Central Flyway; U.S. and Canada population trend –1.2%.

Scissor-tailed Flycatcher: Temperate Nearctic/Nearctic-Neotropic (TG). Variable-distance transitional Neotropic migrant, breeding north to Kansas, and wintering from northern Mexico south to western Panama. Common in the southern Central Flyway; U.S. and Canada population trend –0.7%. (NB). Figure. 38.

Figure 38. Scissor-tailed Flycatcher, adult male.

Family Laniidae: Shrikes

Loggerhead Shrike: Temperate Nearctic/Temperate Nearctic. Variable-distance Nearctic migrant, breeding north to south-central Canada, and wintering from Nebraska south to southern Mexico. Evidence from birds banded in Kansas and reported elsewhere in the Central Flyway states or internationally, or banded elsewhere but reported in Kansas (Table 1) includes movements to Kansas from as far north as Manitoba and South Dakota. One bird, banded in Saskatchewan, was recovered in Oklahoma (Baumgartner and Baumgartner, 1993). Common in the Central Flyway; U.S. and Canada population trend –3.2%. (NB)

Northern Shrike: High-latitude Nearctic/Temperate Nearctic. Variable-distance Holarctic migrant, breeding north to northern Alaska and northwestern Canada, and wintering from southern Canada south to the central Plains States, variably irruptive. Like Snowy Owls, irruptive southward movements in North America average about every 3-5 years, coinciding with population crashes of voles (Newton, 2008). Common in the Central Flyway.

Family Vireonidae: Vireos

Bell's Vireo: Temperate Nearctic/Nearctic–Neotropic. Long-distance Neotropic migrant, breeding north to North Dakota, and wintering from northern Mexico to Nicaragua. As evidence of site-fidelity in a Neotropic migrant, among 248 birds banded in western Nebraska, 2.8 percent were recaptured the following year, 2.4 percent the second year, and 0.8 percent the third year (Scharf *et al*, 2008). Common in the Central Flyway; U.S. and Canada population trend +0.5% (NS). (NB)

Black-capped Vireo: Southern Nearctic: Nearctic-Neotropic. Long-distance Neotropic migrant, breeding north to Oklahoma, and wintering from northwestern Mexico to southern Mexico. Uncommon in the southern Central Flyway. (NB)

Yellow-throated Vireo: Temperate Nearctic/Nearctic-Neotropical (TG). Long-distance transitional Neotropic migrant, breeding north to North Dakota, and wintering from northeastern Mexico south to northern South America, from Suriname to Colombia. Common in the Central Flyway; U.S. and Canada population trend +1.0%.

Plumbeous Vireo: Temperate Nearctic/Nearctic-Neotropic. Variable-distance transitional Neotropic migrant, breeding north to South Dakota and Wyoming, and wintering from western Mexico to northern Central America. Uncommon in the western Central Flyway; U.S. and Canada population trend –1.9%. (NB)

Cassin's Vireo. Extralimital/Nearctic–Neotropic. Long-distance transitional Neotropic migrant, breeding north to central British Columbia, and wintering from northern Mexico to Nicaragua. Rare in the western Central Flyway; U.S. and Canada population trend +1.2%. (NB)

Blue-headed Vireo: Temperate Nearctic/Nearctic–Neotropic (TG). Variable-distance transitional Neotropic migrant, breeding north to southern Northwest Territories, and wintering from the southern U.S, to Central America. Common in the Central Flyway; U.S. and Canada population trend +3.1%. (NB)

Warbling Vireo: Temperate Nearctic/Nearctic-Neotropic (TG). Long-distance transitional Neotropic migrant, breeding north to Canada's Northwest Territories, and wintering from western Mexico to Nicaragua. Evidence from birds banded in Kansas and reported elsewhere in the Central Flyway states or internationally, or banded elsewhere but reported in Kansas (Table 1) includes movements from Kansas to as far as Guatemala and El Salvador. As evidence of site-fidelity in a Neotropic migrant, among 118 birds banded in western Nebraska, 1.7 percent were recaptured the second year after banding, and 1.7 percent the third year (Scharf et al, 2008). A South Dakota-banded bird has been recovered in El Salvador (Tallman, Swanson & Palmer, 2002). Common in the Central Flyway; U.S. and Canada population trend +0.8%. (NB)

Figure 39. Red-eyed Vireo, adult at nest.

Philadelphia Vireo: Temperate Nearctic/Nearctic-Neotropic (TG).
    Long-distance transitional Neotropic migrant, breeding north
    to Canada's Northwest Territories, and wintering from eastern
    Mexico to Colombia. Common in the Central Flyway; U.S. and
    Canada population trend +1.1% (NS).

Red-eyed Vireo: Temperate Nearctic/Central Neotropic (TG). Long-
    distance Neotropic migrant, breeding north to Canada's North-
    west Territories, and wintering in tropical South America, from
    Colombia to northern Argentina. A South Dakota-banded bird
    has been recovered in Costa Rica (Tallman, Swanson & Palmer,
    2002). Common in the Central Flyway; U.S. and Canada popula-
    tion trend +0.9%. Figure 39.

White-eyed Vireo: Temperate Nearctic/Nearctic–Neotropic (TG). Long-distance transitional Neotropic migrant, breeding north to the Great Lakes, and wintering from the southern U.S. to Nicaragua. Common in the Central Flyway; U.S. and Canada population trend +0.6%.

Yellow-green Vireo: Southern Nearctic/Central Neotropic. Long-distance Neotropic migrant, breeding north to southern Texas, and wintering in South America from Colombia to Bolivia. Rare in southernmost Central Flyway. (NB)

Family Corvidae: Jays, Crows. & Relatives

Blue Jay: Temperate Nearctic/Temperate Nearctic. Relatively sedentary Nearctic migrant, but sometimes migratory or widely dispersive. Evidence from birds banded in Kansas and reported elsewhere in the Central Flyway states or internationally, or banded elsewhere but reported in Kansas (Table 1) includes movements to Kansas or from there to as far north as Ontario and Saskatchewan, and as far south as Texas. Of 54 birds banded in Oklahoma, or banded elsewhere but recovered in Oklahoma, the maximum known dispersal distance for an Oklahoma banding documentation was from North Dakota (Baumgartner and Baumgartner, 1993). Common in the Central Flyway. U.S. and Canada population trend +0.7% (NS). Figure 40.

Black-billed Magpie: Temperate Nearctic/Temperate Nearctic. Relative sedentary Nearctic migrant, but somewhat migratory or widely dispersive at northern parts of range. U.S. and Canada population trend +0.5%. Evidence from birds banded in Kansas and reported elsewhere in the Central Flyway states or internationally, or banded elsewhere but reported in Kansas (Table 1) includes a movement from Kansas to Colorado. One bird, banded in Colorado, was recovered there later at a minimum age of five years, seven months. Common in the western Central Flyway. U.S. and Canada population trend +0.5%.

American Crow: Temperate Nearctic/Temperate Nearctic. Relative sedentary Nearctic migrant, but somewhat migratory at north-

Figure 40. Blue Jay, adult.

ern parts of range. Evidence from birds banded in Kansas and reported elsewhere in the Central Flyway states or internationally,
or banded elsewhere but reported in Kansas (Table 1) includes
movements to Kansas or from there to as far north as Alberta,
Saskatchewan and Manitoba, and as far south as Oklahoma. Of
207 birds banded in Oklahoma, or banded elsewhere but recovered in Oklahoma, the maximum known dispersal distance for an
Oklahoma banding documentation was from British Columbia
(Baumgartner and Baumgartner, 1993). One bird, banded in Manitoba, was recovered later in South Dakota at a minimum age of
14 years, seven months. Birds banded elsewhere have been recovered in South Dakota from as far north as Alberta, Saskatchewan
and Manitoba, and as far south as Oklahoma (Tallman, Swanson
& Palmer, 2002). Common in the Central Flyway. U.S. and Canada population trend +0.3%. See Figure 41.

Chihuahuan Raven: Temperate Nearctic/Temperate Nearctic. Relative sedentary Nearctic migrant, but somewhat migratory at
northern parts of range. Evidence from birds banded in Kansas and reported elsewhere in the Central Flyway states or internationally, or banded elsewhere but reported in Kansas (Table
1) includes a movement from Colorado to Kansas. Common in
the southern Central Flyway; U.S. and Canada population trend
+0.3% (NS).

Family Alaudidae: Larks

Horned Lark: Temperate Nearctic/Temperate Nearctic. Variable-distance Holarctic migrant, breeding north to Canada's arctic islands, the more northerly populations the most migratory; the
southern populations are relatively sedentary. Winters from
southern Canada south to central Mexico. One bird, banded in
Colorado, was recovered there later at a minimum age of seven
years, one month. Common in the Central Flyway; U.S. and
Canada population trend –2.2%. See Figure 41.

Figure 41. Horned Lark adult (top), American Crow adult (middle) and Sage Thrasher adult (bottom).

Family Hirundinidae: Swallows & Martins

Purple Martin: Temperate Nearctic/Central Neotropic (TG). Long-distance Neotropic migrant, breeding north to northern Saskatchewan, and wintering in South America, from Suriname to Ecuador and south, east of the Andes, to northern Argentina. Evidence from birds banded in Kansas and reported elsewhere in the Central Flyway states or internationally, or banded elsewhere but reported in Kansas (Table 1) includes movements to Kansas or from there to as far north as Missouri, and as far south as Texas. Of six birds banded in Oklahoma, or banded elsewhere but recovered in Oklahoma, the maximum known dispersal distance for an Oklahoma banding documentation was from Texas (Baumgartner and Baumgartner, 1993). Common in the Central Flyway; U.S. and Canada population trend –0.5%. (NB)

Tree Swallow: Temperate Nearctic/Nearctic-Neotropic (CG). Variable-distance transitional Neotropic migrant, breeding north to northern Alaska and northwestern Canada, and wintering from the southern U.S. to Central America. One bird, banded in Saskatchewan, was recovered there later at a minimum age of nine years. Common in the Central Flyway; U.S. and Canada population trend –1.0%.

Violet-green Swallow: Temperate Nearctic/Nearctic-Neotropic. Variable-distance transitional Neotropic migrant, breeding north to northern Alaska and northwestern Canada, and wintering from northern Mexico to Central America. Uncommon in the western Central Flyway; U.S. and Canada population trend –0.2% (NS). (NB)

Northern Rough-winged Swallow: Temperate Nearctic/Nearctic–Neotropic. Variable-distance transitional Neotropic migrant, breeding north to northern British Columbia, and wintering from the southern U.S. to Panama. As an indication of site-fidelity in a Neotropic migrant, among 230 birds banded in western Nebraska, 0,9 percent were recaptured the following year, and 0.4 percent the third year (Scharf *et al*, 2008). Common in the Central Flyway; U.S. and Canada population trend –0.3% (NS). (NB)

Bank Swallow: Temperate Nearctic/Neotropic (CG). Long-distance Neotropic migrant, breeding north to northern Alaska and northwestern Canada, and wintering through the lowlands of South America south to northern Chile and Argentina. Common in the Central Flyway; U.S. and Canada population trend –2.8% (NS). (NB)

Cliff Swallow: Temperate Nearctic/Neotropic (TG). Long-distance Neotropic migrant, breeding north to northern Alaska and northwestern Canada, and wintering through the lowlands of South America south to northern Argentina and Uruguay. Evidence from birds banded in Kansas and reported elsewhere in the Central Flyway states or internationally, or banded elsewhere but reported in Kansas (Table 1) includes a movement of a banded bird from Kansas to Argentina. A South Dakota-banded bird was recovered in West Virginia (Tallman, Swanson & Palmer, 2002). One bird, banded in Nebraska, was recovered later in California at a minimum age of 11 years, ten months. Brown and Brown (1996) reported that, on average about 59 percent of the Cliff Swallows in their Nebraska colonies returned (after migrating to and wintering in tropical South America) to breed in their previous-year colony site, for periods of at least up to six years! Two female Cliff Swallows, banded as nestlings at Cedar Point, are known to have survived at least 13 years, based on recaptures there (Mary B. Brown, pers. comm.), presumably having traveled about 100,000 miles during their 12 round-trips of 8,000 mile each to and from equatorial wintering areas. These miles do not include the uncountable number of miles covered while foraging during most hours of daylight every day that they are not migrating. Common in the Central Flyway; U.S. and Canada population trend +0.8%. (NB)

Cave Swallow: Southern Nearctic/Nearctic-Neotropic. Variable-distance Neotropic migrant, breeding north to northern Texas, and probably wintering to southern Mexico, but wintering range unclear. Uncommon in the southern Central Flyway; U.S. and Canada population trend +11.8%. (NB)

Barn Swallow: Temperate Nearctic/Neotropic (TG). Variable-distance Neotropic migrant, breeding north to southern Alaska and Northwest Territories in Canada; wintering from Mexico throughout the lowlands of Central and South America. Evidence from birds banded in Kansas and reported elsewhere in the Central Flyway states or internationally, or banded elsewhere but reported in Kansas (Table 1) includes a movement of a banded bird from Kansas to Guatemala. One bird, banded in Oklahoma, was recovered in Bolivia (Baumgartner and Baumgartner, 1993). Common in the Central Flyway; U.S. and Canada population trend –1.2%. (NB)

Family Paridae: Chickadees and Tits

Carolina Chickadee: Temperate Nearctic/Temperate Nearctic. Relative sedentary Nearctic migrant, but somewhat migratory at northern parts of range. Common in the Central Flyway. U.S. and Canada population trend –0.1% (NS).

Black-capped Chickadee: Temperate Nearctic/Temperate Nearctic. Relative sedentary Nearctic migrant, but sometimes migratory and/or irruptive at northern parts of range. Largely dependent during winter on conifer seeds, and thus considered an irruptive species (Newton, 2008). One bird, banded in Minnesota, was recovered there later at a minimum age of 11 years, six months. As evidence of well-developed site-fidelity in a relatively sedentary passerine, of 271 birds banded in western Nebraska, 4.4 percent were recaptured the following year, 2.2 percent the second year, 0.7 percent during each of the third year and fourth years, and 0.4 percent the fifth year (Scharf *et al*, 2008). Common in the southern Central Flyway; U.S. and Canada population trend +1.0%.

Family Sittidae: Nuthatches

Red-breasted Nuthatch: Temperate Nearctic/Temperate Nearctic. Variable-distance Nearctic migrant, breeding north to southern Alaska. Only the most northerly populations are regularly migratory; many populations are relatively irruptive in the extent of their winter movements, depending on the availability of pine seeds (Newton, 2008). Winters from southern Alaska and the southern Canada south to central Mexico. Common in the Central Flyway; U.S. and Canada population trend +1.7%.

White-breasted Nuthatch: Nearctic/Temperate Nearctic. Variable-distance Nearctic migrant, breeding north to central Alberta and Saskatchewan. Only the most northerly populations are regularly migratory, but the southern pupations are prone to wander, at least in Texas and Oklahoma, and in eastern Kansas there is a fall arrival of migrants, with birds arriving in August and September. Common in the Central Flyway; U.S. and Canada population trend +2.0%. Figure 42.

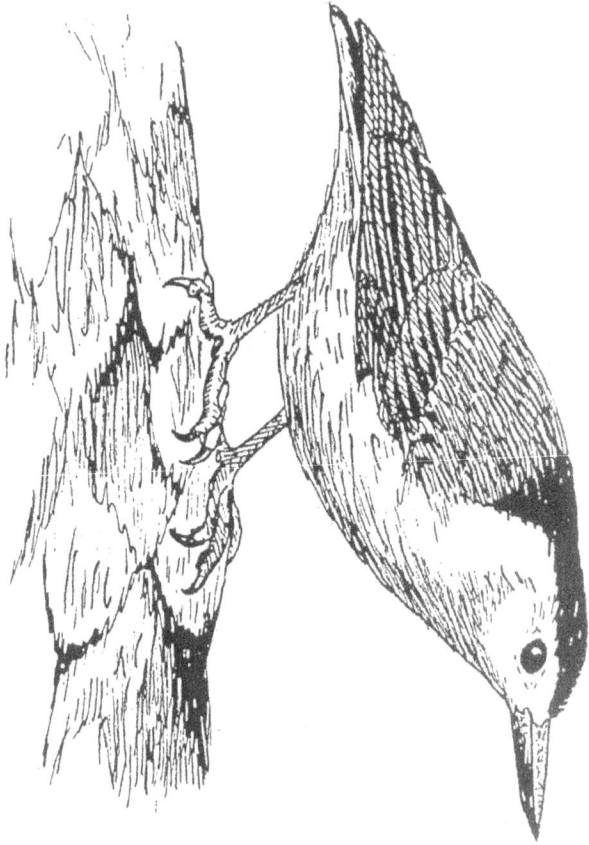

Figure 42. White-breasted Nuthatch, adult.

Family Certhiidae: Creepers

Brown Creeper: Temperate Nearctic. Variable-distance Nearctic migrant, breeding north to southern Alaska, where sedentary, the more northerly populations of interior Canada are the most migratory. Winters from southern Canada south to Texas. Common in the Central Flyway; U.S. and Canada population trend +0.9%. (NB)

Family Troglodytidae: Wrens

Rock Wren: Temperate Nearctic/Temperate Nearctic. Variable-distance Nearctic migrant, breeding north to southern British Columbia, the more northerly populations the most migratory; the southern populations are relatively sedentary. Winters from Washington south to New Mexico. Common in the western Central Flyway; U.S. and Canada population trend –0.8%.

Carolina Wren: Temperate Nearctic/Temperate Nearctic. Relative sedentary Nearctic migrant, but somewhat migratory at northern parts of range. Common in the southern Central Flyway; U.S. and Canada population trend +1.4%.

House Wren: Temperate Nearctic/Nearctic–Neotropic (TG). Variable-distance transitional Neotropic migrant, breeding north to Canada's Northwest Territories, all but the most southerly populations are migratory. Winters from Oklahoma south to southern Mexico. One bird, banded in Oklahoma, was recovered in Illinois (Baumgartner and Baumgartner, 1993). Birds banded elsewhere have been recovered in South Dakota from as far away as Missouri (Tallman, Swanson & Palmer, 2002). As evidence of site-fidelity in a transitional Neotropic migrant, among 416 birds banded in western Nebraska, 5.5 percent were recaptured the following year, 1.0 percent the second year, and 0.2 percent during each of the third and fourth years (Scharf et al, 2008). Common in the Central Flyway; U.S. and Canada population trend +1.3%.

Winter Wren: Temperate Nearctic/Temperate Nearctic. Variable-distance Nearctic migrant, breeding north to central British Columbia, all but the most southerly populations are migratory. Winters from the Ohio Valley south to coastal Texas. Common in the Central Flyway; U.S. and Canada population trend +1.1%.

Sedge Wren: Temperate Nearctic/Temperate Nearctic. Short-distance Nearctic migrant, breeding north to central Alberta and Saskatchewan, all breeding populations are migratory. Winters from Arkansas and Tennessee south to northeastern Mexico. Common in the Central Flyway; U.S. and Canada population trend +1.7% (NS). (NB)

Figure 43. Marsh Wren, adult.

Marsh Wren: Temperate Nearctic/Nearctic–Neotropic (TG). Variable-distance transitional Neotropic migrant, breeding north to northern Alberta, the more northerly populations the most migratory; the southern populations are relatively sedentary. Winters in the central Plains from Colorado south to southern Mexico. Common in the Central Flyway; U.S. and Canada population trend +2.1%. (NB). Figure 43.

Family Polioptilidae: Gnatcatchers

Blue-gray Gnatcatcher: Temperate Nearctic/Nearctic–Neotropic (CG) Variable-distance transitional Neotropic migrant, breeding north to central Minnesota and Maine, most breeding populations are migratory; the Gulf Coast and Southeastern states' populations are relatively sedentary. Winters from the Gulf and South Atlantic states south to Honduras. Common in the Central Flyway; U.S. and Canada population trend +0.8%. (NB)

Family Regulidae: Kinglets

Golden-crowned Kinglet: Temperate Nearctic/Temperate Nearctic. Variable-distance Nearctic migrant, breeding north to southern Alaska, the more northerly populations the most migratory; the southern populations are relatively sedentary. Winters in the central Plains from South Dakota south to southern Texas and northeastern Mexico. One bird, banded in Minnesota, was recovered there later at a minimum age of six years, four months. Common in the Central Flyway; U.S. and Canada population trend –0.6% (NS). (NB)

Ruby-crowned Kinglet: Temperate Nearctic/Nearctic–Neotropic (CG), Variable-distance transitional Neotropic migrant, breeding north to northern Alaska, nearly all populations are migratory. Winters in the central Plains from Oklahoma south to southern Mexico. One bird, banded in Illinois, was recovered in Oklahoma (Baumgartner and Baumgartner, 1993). Common in the Central Flyway; U.S. and Canada population trend +0.1% (NS). (NB)

Family Turdidae: Thrushes & Allies

Eastern Bluebird: Temperate Nearctic/Temperate Nearctic. Variable-distance Nearctic migrant, breeding north to central Ontario, the more northerly populations the most migratory; the southern populations are relatively sedentary. Winters in the central Plains from South Dakota south to northeastern Mexico. Evidence from birds banded in Kansas and reported elsewhere in the Central Flyway states or internationally, or banded elsewhere but reported in Kansas (Table 1) includes movements to Kansas or from there to as far north as Manitoba and as far south as Texas. One bird, banded in Manitoba, was recovered in Oklahoma (Baumgartner and Baumgartner, 1993). Common in the Central Flyway; U.S. and Canada population trend +1.9%. Figure 44.

Mountain Bluebird: Temperate Nearctic/Temperate Nearctic. Variable-distance Nearctic migrant, breeding north to central Alaska, the more northerly populations the most migratory; the southern populations are relatively sedentary. Winters in the central Plains from western Kansas south to western Texas and northern Mexico. Evidence from birds banded in Kansas and reported elsewhere in the Central Flyway states or internationally, or banded elsewhere but reported in Kansas (Table 1) includes movements to Kansas or from there to as far north as Alberta. One bird, banded in Alberta, was recovered there later at a minimum age of nine years. Uncommon in the western Central Flyway; U.S. and Canada population trend –0.7%.

Western Bluebird: Temperate Nearctic/Temperate Nearctic. Variable-distance Nearctic migrant, breeding north to central British Columbia, the more northerly populations the most migratory; the southern populations are relatively sedentary. Winters in the central Plains from southern Colorado south to central Mexico. Uncommon to rare in the western Central Flyway; U.S. and Canada population trend –0.7% (NS)

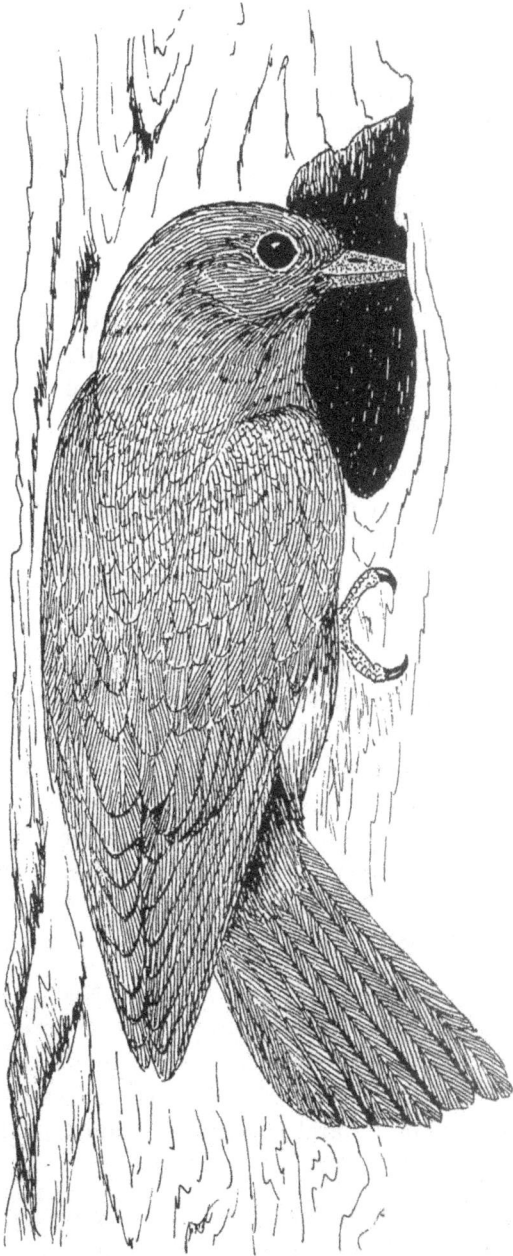

Figure 44. Eastern Bluebird, adult at nest.

Townsend's Solitaire: Temperate Nearctic/Temperate Nearctic. Variable-distance Nearctic migrant, breeding north to northern Alaska, northern Yukon Territory and Northwest Territories, the more northerly populations the most migratory; the southern populations are relatively sedentary. Winters in the Great Plains from Alberta and Saskatchewan south to northern Mexico. Uncommon in the western Central Flyway; U.S. and Canada population trend +0.8% (NS). (NB)

Veery: Temperate Nearctic/Neotropic (TG). Long-distance Neotropic migrant, breeding north to northern Alberta and central Saskatchewan. Winters in South America from Colombia to Bolivia and western Brazil. Common in the Central Flyway; U.S. and Canada population trend +0.8%.

Gray-cheeked Thrush: Temperate Nearctic/Neotropic (TG). Long-distance Neotropic migrant, breeding north to northern Alaska, northern Yukon Territory and Northwest Territories. Winters in South America from Colombia to Bolivia and northwestern Brazil. Common in the Central Flyway.

Swainson's Thrush: Temperate Nearctic/Neotropic (TG). Long-distance Neotropic migrant, breeding north to northern Alaska and northwestern Canada. Winters in South America from Venezuela to Ecuador and south to northern Argentina. One bird, banded in Montana, was recovered there later at a minimum age of 12 years, one month. A South Dakota-banded bird was recovered in Guatemala (Tallman, Swanson & Palmer, 2002). Common in the Central Flyway; U.S. and Canada population trend +0.8% (NS).

Hermit Thrush: Temperate Nearctic/Nearctic-Neotropic (TG). Variable-distance transitional Neotropic migrant, breeding north to northern Alaska and northern Northwest Territories. Winters in the Great Plains from Oklahoma south to southern Mexico. Common in the Central Flyway; U.S. and Canada population trend +0.9% (NS).

Wood Thrush: Temperate Nearctic/Nearctic-Neotropic (TG). Long-distance transitional Neotropic migrant, breeding north to northern Minnesota and the southern Canada. Winters from southeastern Mexico to Colombia. One bird, banded in Illinois, was recovered in Oklahoma (Baumgartner and Baumgartner, 1993). Common in the western Central Flyway; U.S. and Canada population trend –1.8%.

American Robin: Temperate Nearctic/Temperate Nearctic. Variable-distance transitional Neotropic migrant, breeding north to northern Alaska, Yukon Territory, and northern Northwest Territories; most populations are migratory; the southeastern populations are relatively sedentary. Winters in the southern U.S. from the Gulf Coast south to southern Mexico. Evidence from birds banded in Kansas and reported elsewhere in the Central Flyway states or internationally, or banded elsewhere but reported in Kansas (Table 1) includes movements to Kansas or from there to as far north as Alaska and as far south as Texas. Of 25 birds banded in Oklahoma, or banded elsewhere but recovered in Oklahoma, the maximum known dispersal distance for an Oklahoma banding documentation was from Pennsylvania (Baumgartner and Baumgartner, 1993). South Dakota-banded birds have been recovered south to Texas (about half of 478 total recoveries) and north to Saskatchewan, as well as east to North Carolina and Florida (Tallman, Swanson & Palmer, 2002). As evidence of site-fidelity in a relatively short-distance migrant, among 144 birds banded in western Nebraska, 6.3 percent were recaptured the following year, and 2.7 percent the second year (Scharf et al, 2008). Common in the Central Flyway; U.S. and Canada population trend +0.3%. (NB)

Family Mimidae: Mockingbirds, Thrashers & Catbirds

Gray Catbird: Temperate Nearctic/Nearctic-Neotropic (TG). Variable-distance transitional Neotropic migrant, breeding north to northern Alberta; most populations are migratory; the southeastern populations are relatively sedentary. A South Dakota-banded bird was recovered in Saskatchewan (Tallman, Swanson & Palmer, 2002). Winters from the Gulf Coast south to Panama.

Common in the Central Flyway; U.S. and Canada population trend –0.1% (NS).

Northern Mockingbird: Temperate Nearctic/Temperate Nearctic. Variable-distance Nearctic migrant, breeding in the Great Plains north to Nebraska; more northern populations are migratory; the southern populations are relatively sedentary. Winters from Kansas south to southern Mexico. One bird, banded in Texas, was recovered there later at a minimum age of 14 years, ten months. Of three birds banded in Oklahoma, the maximum known dispersal distance for a recovery was from Texas (Baumgartner and Baumgartner, 1993). As evidence of site-fidelity in a Neotropic migrant, among 144 birds banded in western Nebraska, 6.3 percent were recaptured the following year, and 2.8 percent the second year (Scharf et al, 2008). Common in the southern Central Flyway; U.S. and Canada population trend –0.5%.

Sage Thrasher: Temperate Nearctic/Temperate Nearctic. Variable-distance Nearctic migrant, breeding in the western plains north to northern Montana. Winters from New Mexico south to central Mexico. Common in the western Central Flyway; U.S. and Canada population trend –0.6% (NS). See Figure 41.

Brown Thrasher: Temperate Nearctic/Temperate Nearctic. Variable-distance Nearctic migrant, breeding in the Great Plains north to central Alberta and Saskatchewan; northern populations are migratory; the southern populations are relatively sedentary. Winters from Oklahoma south to southern Texas. Evidence from birds banded in Kansas and reported elsewhere in the Central Flyway states or internationally, or banded elsewhere but reported in Kansas (Table 1) includes movements to Kansas or from there to as far north as Nebraska and as far south as Texas. Of six birds banded in Oklahoma, or banded elsewhere but recovered in Oklahoma, the maximum known dispersal distance for a banded bird was from Illinois (Baumgartner and Baumgartner, 1993). Birds banded elsewhere have been recovered in South Dakota from as far away as Illinois, and South Dakota-banded birds have been recovered from North Dakota to Texas (Tallman, Swanson & Palmer, 2002). Common in the southern Central Flyway; U.S. and Canada population trend –0.3% (NS).

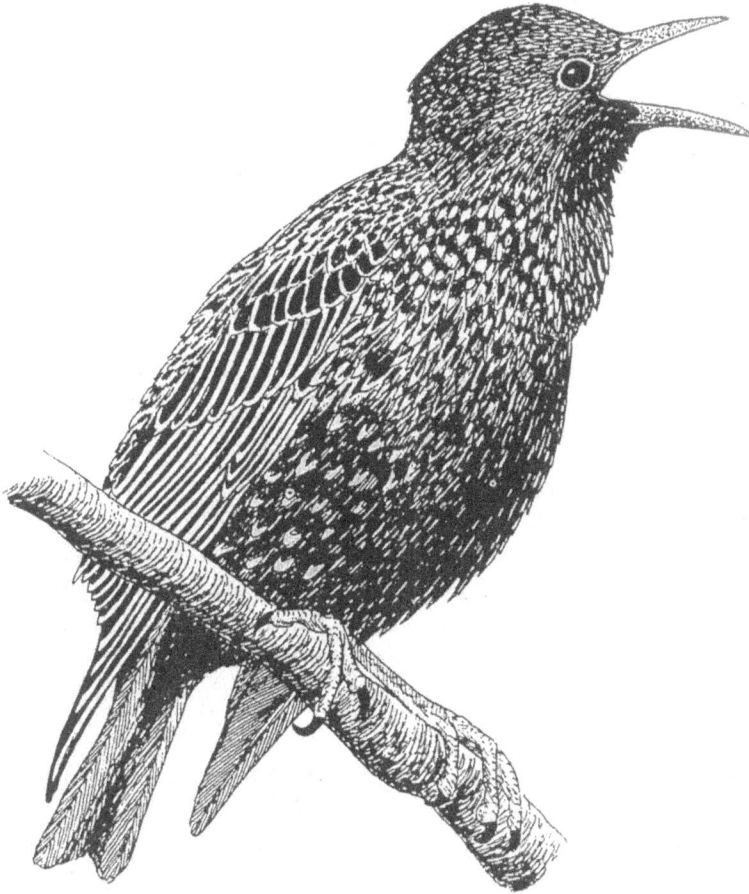

Figure 45. European Starling, adult.

Family Sturnidae: Starlings

European Starling: Temperate Nearctic/Temperate Nearctic. Variable-distance Holarctic migrant, breeding in the Great Plains north to southern Alaska and the southern Northwest Territories; north-ernmost populations are migratory; the other populations are relatively sedentary. Winters from southern Canada south to northern Mexico. Evidence from birds banded in Kansas and reported elsewhere in the Central Flyway states or internationally, or banded elsewhere but reported in Kansas (Table 1) includes encounters to Kansas or from there to as far north as Minnesota and as far south as Texas. Of 35 birds banded in Oklahoma, or banded elsewhere but recovered in Oklahoma, the maximum known dispersal distance for a banded bird was from Minnesota (Baumgartner and Baumgartner, 1993). Birds banded elsewhere have been recovered in South Dakota from as far away as Utah and Colorado, and South Dakota-banded birds have been recovered from North Dakota to Oklahoma (Tallman, Swanson & Palmer, 2002). Abundant in the Central Flyway; U.S. and Canada population trend –1.2%. (NB). Figure 45.

Family Motacillidae: Pipits

American Pipit: High-latitude Nearctic/Nearctic–Neotropic (CG), Variable-distance Nearctic migrant, breeding north to northernmost Alaska and Canada's arctic islands; all populations (arctic- and alpine-breeding) migratory; Winters in the Great Plains from Oklahoma south to southern Mexico. Common in the Central Flyway.

Sprague's Pipit: Temperate Nearctic/Nearctic–Neotropic. Long-distance transitional Neotropic migrant, breeding in the Great Plains north to central Alberta and Saskatchewan, Winters from Oklahoma south to southern Mexico. Common in the Central Flyway; U.S. and Canada population trend –2.2%.

Family Bombycillidae: Waxwings

Bohemian Waxwing: High-latitude Nearctic/Temperate Nearctic. Variable-distance Holarctic migrant, breeding north to northern Alaska and northwestern Canada; northern populations are migratory and/or irruptive; the southern populations are relatively sedentary. Winters from northern Alberta and northern Saskatchewan south to the Dakotas and sometimes to northern Nebraska, with movements somewhat irruptive from year to year. One bird, banded in Saskatchewan, was recovered later in British Columbia at a minimum age of five years, ten months. Birds banded elsewhere have been recovered in South Dakota from as far away as British Columbia (Tallman, Swanson & Palmer, 2002). Widespread in the northern Central Flyway.

Cedar Waxwing: Temperate Nearctic/Nearctic–Neotropic (TG), Variable-distance transitional Neotropic migrant, breeding north to southeastern Alaska, northern Alberta and northern Saskatchewan; northern populations are migratory; the southern populations are relatively sedentary. Winters from North Dakota south to Panama, rarely to Venezuela and Colombia. Evidence from birds banded in Kansas and reported elsewhere in the Central Flyway states or internationally, or banded elsewhere but reported in Kansas (Table 1) includes movements to Kansas or from there to as far north as Minnesota. One bird, banded in Minnesota, was recovered in Oklahoma (Baumgartner and Baumgartner, 1993). South Dakota-banded birds have been recovered in Texas, Ohio and Minnesota (Tallman, Swanson & Palmer, 2002). Common in the Central Flyway; U.S. and Canada population trend +0.3% (NS).

Family Calcariidae: Longspurs & Snow Buntings

Lapland Longspur: High-latitude Nearctic/Temperate Nearctic. Long-distance Holarctic migrant, breeding north to northernmost Alaska and Canada's arctic islands; all populations are migratory. Winters in the Great Plains from Manitoba and Saskatchewan south to Texas. Common in the northern Central Flyway.

Chestnut-collared Longspur: Temperate Nearctic/Southern Nearctic. Short-distance Nearctic migrant, breeding in the Great Plains north to Alberta and Saskatchewan; all populations are migratory. Winters in the Great Plains from Oklahoma to Texas, and northern Mexico. Common in the Central Flyway, and a Central Flyway endemic; U.S. and Canada population trend –4.3%.

Smith's Longspur: High-latitude Nearctic/Southern Nearctic. Long-distance Nearctic migrant, breeding north to northern Alaska and northern Canada; all populations are migratory. Winters in the Great Plains from southern Kansas to northern Texas. Common in the southern Central Flyway.

McCown's Longspur: Temperate Nearctic/Southern Nearctic. Short-distance Nearctic migrant, breeding in the Great Plains north to Alberta and Saskatchewan; all populations are migratory. Winters in the Great Plains from Oklahoma to Texas and northern Mexico. Common in the western Central Flyway and a Central Flyway endemic;.

Snow Bunting: High-latitude Nearctic/Temperate Nearctic. Long-distance Holarctic migrant, breeding north to northernmost Alaska and Canada's arctic islands; all populations are migratory. Winters in the Great Plains from Manitoba and Saskatchewan south often to Nebraska, the distances varying with winter severity. Common in the northern Central Flyway.

Family Parulidae: Wood Warblers

Ovenbird: Temperate Nearctic/Nearctic–Neotropic (TG). Long-distance transitional Neotropic migrant, breeding north to southern Northwest Territories; all populations are migratory. Winters from northern Mexico to northern South America, from Venezuela to Colombia, and in the Caribbean region. Evidence from birds banded in Kansas and reported elsewhere in the Central Flyway states or internationally, or banded elsewhere but reported in Kansas (Table 1) includes a banded bird record in Kansas of a bird banded in Minnesota the previous spring. Common in the Central Flyway; U.S. and Canada population trend +0.1% (NS).

Worm-eating Warbler: Temperate Nearctic/Nearctic–Neotropic (TG). Long-distance transitional Neotropic migrant, breeding north to southern Wisconsin; all populations are migratory. Winters from east-central Mexico to Panama, and in the Caribbean region. Uncommon in southeastern Central Flyway; U.S. and Canada population trend +0.6% (NS).

Louisiana Waterthrush: Temperate Nearctic/Nearctic–Neotropic (TG). Long-distance transitional Neotropic migrant, breeding north to central Minnesota; all populations are migratory. Winters from northern Mexico to northern South America, from Venezuela to Colombia, and in the Caribbean region. Common in southeastern Central Flyway; U.S. and Canada population trend +0.4% (NS).

Northern Waterthrush: Temperate Nearctic/Central Neotropic (TG). Long-distance Neotropic migrant, breeding north to northern Alaska and northern Northwest Territories; all populations are migratory. Winters from northern Mexico to northern South America, from Ecuador to Peru, and in the Caribbean region. Common in the eastern Central Flyway; U.S. and Canada population trend +0.6% (NS).

Golden-winged Warbler: Temperate Nearctic/Neotropic (TG). Long-distance Neotropic migrant, breeding north to Minnesota; all populations are migratory. Winters in South America, from Venezuela to Colombia. Uncommon in the eastern Central Flyway; U.S. and Canada population trend –2.5%.

Black-and-White Warbler: Temperate Nearctic/Nearctic–Neotropic (TG). Long-distance transitional Neotropic migrant, breeding north to central Northwest Territories; all populations are migratory. Winters from northern Mexico to northern South America, from Venezuela to Peru, and in the Caribbean region. Evidence from birds banded in Kansas and reported elsewhere in the Central Flyway states or internationally, or banded elsewhere but reported in Kansas (Table 1) includes a banded bird in Sonora, Mexico, of a Kansas-banded bird. Common in the Central Flyway; U.S. and Canada population trend +0.2% (NS). Figure 46.

Prothonotary Warbler: Temperate Nearctic/Neotropic (TG). Long-distance Neotropic migrant, breeding north to central Minnesota; all populations are migratory. Winters from southeastern

Figure 46. Black-and-White Warbler, adult male.

Mexico to northern South America, from Guyana to Ecuador. Uncommon in southeastern Central Flyway; U.S. and Canada population trend –0.9%.

Swainson's Warbler: Temperate Nearctic/Nearctic–Neotropic (TG). Long-distance Neotropic migrant, breeding north to Missouri and West Virginia; all populations are migratory. Winters from east-central Mexico to Belize. Uncommon in southeastern Central Flyway; U.S. and Canada population trend +1.1% (NS).

Tennessee Warbler: Temperate Nearctic/Neotropic (TG) Long-distance Neotropic migrant, breeding north to northern Northwest Territories; all populations are migratory. Birds banded elsewhere have been recovered in South Dakota from as far away as Michigan (Tallman, Swanson & Palmer, 2002). Winters from southern Mexico to northern South America, from Venezuela to Colombia. Common in the Central Flyway; U.S. and Canada population trend –0.4% (NS).

Orange-crowned Warbler: Temperate Nearctic/Nearctic–Neotropic (CG), Variable-distance transitional Neotropic migrant, breeding north to northern Alaska and northern Northwest Territories; nearly all populations are migratory. Winters from coast of Pacific Northwest to Guatemala. Evidence from birds banded in Kansas and reported elsewhere in the Central Flyway states or internationally, or banded elsewhere but reported in Kansas (Table 1) includes the movement of a Kansas-banded bird to Michoacan, Mexico. Birds banded elsewhere have been recovered in South Dakota from as far away as Alaska, and a South Dakota-banded bird was recovered in Wisconsin (Tallman, Swanson & Palmer, 2002). Common in the Central Flyway; U.S. and Canada population trend –0.9%. (NB)

Colima Warbler: Southern Nearctic/Nearctic–Neotropic. Short-distance transitional Neotropic migrant, breeding north to Big Bend area of southwestern Texas; all populations are migratory. Winters in west-central Mexico. Local and rare in southwestern Central Flyway.

Lucy's Warbler: Temperate Nearctic/Nearctic–Neotropic. Short-distance transitional Neotropic migrant, breeding north to western Colorado; all populations are migratory. Winters in western Mexico. Local (western Rio Grande Valley) in the southwestern Central Flyway; U.S. and Canada population trend +0.3%. (NB)

Nashville Warbler: Temperate Nearctic/Nearctic–Neotropic (TG). Long-distance transitional Neotropic migrant, breeding north to northern Alberta; all populations are migratory. Winters from southern Texas and northwestern Mexico to Guatemala. Evidence from birds banded in Kansas and reported elsewhere in the Central Flyway states or internationally, or banded elsewhere but reported in Kansas (Table 1) includes the movement of a Kansas-banded bird to Minnesota. One bird, banded in Oklahoma, was recovered in Mexico (Baumgartner and Baumgartner, 1993). Common in the Central Flyway; U.S. and Canada population trend +0.1%.

Virginia's Warbler: Temperate Nearctic/Nearctic–Neotropic. Long-distance transitional Neotropic migrant, breeding north to central Wyoming; all populations are migratory. Winters in western Mexico. Local (western Texas) in the southern Central Flyway; U.S. and Canada population trend –0.8% (NS).

Connecticut Warbler: Temperate Nearctic/Central Neotropic (TG). Long-distance Neotropic migrant, breeding north to northern British Columbia; all populations are migratory. Migrates to northern South America, probably via a Western Atlantic overseas flight, from the U.S. East Coast south to coastal South America. Winters in northern South America, including Venezuela, Colombia and west-central Brazil. Uncommon in the northern Central Flyway; U.S. and Canada population trend –1.0% (NS).

MacGillivray's Warbler: Temperate Nearctic/Nearctic–Neotropic. Long-distance transitional Neotropic migrant, breeding north to southern Alaska and Northwest Territories; all populations are migratory. Winters from northwestern Mexico to western Panama. One bird, banded in Wyoming, was recovered there later at a minimum age of nine years. Common in the western Central Flyway; U.S. and Canada population trend –0.6%.

Mourning Warbler: Temperate Nearctic/Central Neotropic (TG/CG). Long-distance Neotropic migrant, breeding north to southern Northwest Territories; all populations are migratory. Winters from southern Central America south to Venezuela and Ecuador. Common in the Central Flyway; U.S. and Canada population trend –1.4%.

Kentucky Warbler: Temperate Nearctic/Nearctic–Neotropic (TG). Long-distance transitional Neotropic migrant, breeding north to northern Iowa; all populations are migratory. Winters from northwestern Mexico to western Panama. Common in the eastern Central Flyway; U.S. and Canada population trend –0.7%.

Common Yellowthroat: Temperate Nearctic/Nearctic–Neotropic (TG). Variable-distance transitional Neotropic migrant, breeding north to central Yukon and Northwest Territories; nearly all populations are migratory. Winters from the southern U.S. to Venezuela and Colombia, and in the Caribbean region. As evidence of site-fidelity in a Neotropic migrant, among 329 birds banded in western Nebraska, 1.5 percent were recaptured the following year, 0,9 percent the second year, 0.9 percent during the third year, and 0.3 percent the fourth year (Scharf et al, 2008). Common in the Central Flyway; U.S. and Canada population trend –0.9%. (NB)

Hooded Warbler: Temperate Nearctic/Neotropic (TG). Long-distance Neotropic migrant, breeding north to central Wisconsin; all populations are migratory. Winters from southeastern Mexico to Panama, and in the Caribbean region. Uncommon in the southern Central Flyway; U.S. and Canada population trend +1.7%.

American Redstart: Temperate Nearctic/Central Neotropic (TG). Long-distance Neotropic migrant, breeding north to northern Northwest Territories; all populations are migratory. Winters from northern Mexico to Ecuador and Brazil, and the Caribbean region. Common in the Central Flyway; U.S. and Canada population trend –0.4% (NS).

Cape May Warbler: Temperate Nearctic/Neotropic (TG). Long-distance Neotropic migrant, breeding north to central Northwest

Territories; all populations are migratory. Winters from Yucatan, Mexico south through Central America, and in the Caribbean region. Common in the Central Flyway; U.S. and Canada population trend –2.8%.

Cerulean Warbler: Temperate Nearctic/Central Neotropic (TG). Long-distance Neotropic migrant, breeding north to central Minnesota; all populations are migratory. Winters in South America, from Venezuela and Colombia south to Bolivia. Uncommon to rare in the eastern Central Flyway; U.S. and Canada population trend –2.9%.

Northern Parula: Temperate Nearctic/Nearctic–Neotropic (TG). Long-distance transitional Neotropic migrant, breeding north to southern Manitoba; nearly all populations are migratory. Winters from northern Mexico south to northern Central America, and in the Caribbean region. Common in the eastern Central Flyway; U.S. and Canada population trend +1.1%.

Magnolia Warbler: Temperate Nearctic/Nearctic–Neotropic (TG). Long-distance transitional Neotropic migrant, breeding north to central Northwest Territories; all populations are migratory. Winters in the Caribbean region and from central Mexico south to Panama. Common in the eastern Central Flyway; U.S. and Canada population trend +0.4% (NS).

Bay-breasted Warbler: Temperate Nearctic/Central Neotropic (TG). Long-distance Neotropic migrant, breeding north to southern Manitoba; all populations are migratory. Winters from Panama to Venezuela. One bird, banded in Iowa, was recovered later in Panama at a minimum age of three years, five months. Common in the eastern Central Flyway; U.S. and Canada population trend –0.3% (NS).

Blackburnian Warbler: Temperate Nearctic/Central Neotropic (TG). Long-distance Neotropic migrant, breeding north to central Manitoba and Saskatchewan; all populations are migratory. Winters in South America, from Venezuela and Colombia south to Bolivia. One bird, banded in Minnesota, was recovered there later at a minimum age of eight years, two months. Common in the eastern Central Flyway; U.S. and Canada population trend +0.7% (NS).

Yellow Warbler: Temperate Nearctic/Nearctic–Neotropic (TG). Long-distance transitional Neotropic migrant, breeding north to northern Alaska and northern Northwest Territories; all populations are migratory. Winters from the southern U.S. south through Central America to Peru and Brazil, and in the Caribbean region. One bird, banded in Colorado, was recovered later in Nebraska at a minimum age of nine years, 11 months. As evidence of site-fidelity in a Neotropic migrant, among 857 birds banded in western Nebraska, 2.0 percent were recaptured the following year, 1.5 percent the second year, and 0.7 percent during the third year (Scharf et al, 2008). Common in the Central Flyway; U.S. and Canada population trend –0.4%. (NB)

Chestnut-sided Warbler: Temperate Nearctic/Neotropic (TG). Long-distance Neotropic migrant, breeding north to central Alberta; all populations are migratory. Winters from southern Mexico, south to northern South America, from Venezuela and Colombia south to Bolivia, and in the Caribbean region. Common in the eastern Central Flyway; U.S. and Canada population trend +1.4%.

Blackpoll Warbler: Temperate Nearctic/Central Neotropic (TG). Long-distance Neotropic migrant, breeding north to northern Alaska and northern Northwest Territories; all populations are migratory. Winters from southern Mexico south to northern South America, from Venezuela and Colombia south to Bolivia, and in the Caribbean region. It is likely that this species regularly migrates non-stop over the Western Atlantic Ocean from the coast of New England to the northern coast of South America (Nisbet, 1970, Ralph, 1978, Bildstein, 1999). Common in the eastern Central Flyway; U.S. and Canada population trend –1.0%.

Black-throated Blue Warbler: Temperate Nearctic/Neotropic (TG). Long-distance Neotropic migrant, breeding north to southern Ontario; all populations are migratory. Winters from southern Mexico south to northern South America, in northern Venezuela and northern Colombia, and also in the Caribbean region. Uncommon in the eastern Central Flyway; U.S. and Canada population trend +2.2%.

Palm Warbler: Temperate Nearctic/Neotropic (TG). Long-distance Neotropic migrant, breeding north to northern Northwest Territories; all populations are migratory. Winters from southern Mexico south to Honduras, and in the Caribbean region. Uncommon in the eastern Central Flyway; U.S. and Canada population trend −2.0% (NS).

Pine Warbler: Temperate Nearctic/Southern Nearctic. Variable-distance Nearctic migrant, breeding north to northern Northwest Territories; northern populations are migratory; the southern populations are relatively sedentary. Winters from Oklahoma to southern Texas, and in the Caribbean region. Uncommon in the southern Central Flyway; U.S. and Canada population trend +1.3%.

Yellow-rumped Warbler: Temperate Nearctic/Nearctic–Neotropic (TG). Variable-distance transitional Neotropic migrant, breeding north to northern Alaska and northern Northwest Territories; nearly all populations are migratory. Winters from Kansas south to Honduras, and in the Caribbean region. One bird, banded in Wyoming, was recovered there later at a minimum age of ten years. One bird, banded in North Dakota, was recovered in Oklahoma (Baumgartner and Baumgartner, 1993). A South Dakota-banded bird was recovered in Louisiana (Tallman, Swanson & Palmer, 2002). Common in the Central Flyway; U.S. and Canada population trend −0.1% (NS). (NB)

Yellow-throated Warbler: Temperate Nearctic/Nearctic–Neotropic (TG). Long-distance transitional Neotropic migrant, breeding north to southern Wisconsin; nearly all populations are migratory. Winters from the U.S. and Mexican Gulf Coast south to Central America, and in the Caribbean region. Common in the southern Central Flyway; U.S. and Canada population trend +1.0%. (NB)

Prairie Warbler: Temperate Nearctic/Nearctic–Neotropic (TG). Long-distance transitional Neotropic migrant, breeding north to Missouri; nearly all populations are migratory. Winters from the Yucatan coast south to Central America, and in the Caribbean region. Common in the southern Central Flyway; U.S. and Canada population trend −2.2%.

Grace's Warbler: Temperate Nearctic/Nearctic–Neotropic. Long-distance transitional Neotropic migrant, breeding north to western Colorado; U.S. populations are migratory. Winters from west-central Mexico south to Nicaragua. Common in the southern Central Flyway; U.S. and Canada population trend –1.5% (NS).

Black-throated Gray Warbler: Temperate Nearctic/Nearctic–Neotropic. Long-distance transitional Neotropic migrant, breeding north to southern British Columbia; all populations are migratory. Winters from northwestern Mexico south to southern Mexico. Rare in southwestern Central Flyway; U.S. and Canada population trend –0.9%.

Townsend's Warbler: Temperate Nearctic/Nearctic–Neotropic. Variable-distance transitional Neotropic migrant, breeding north to central Alaska; all populations are migratory. Winters from Puget Sound south to Costa Rica. Rare in the western Central Flyway; U.S. and Canada population trend +0.8% (NS).

Golden-cheeked Warbler: Temperate Nearctic/Neotropic. Long-distance Neotropic migrant, breeding locally in central Texas; all populations are migratory. Winters from southern Mexico to Nicaragua. Rare in extreme southern Central Flyway.

Black-throated Green Warbler: Temperate Nearctic/Nearctic–Neotropic (TG). Long-distance transitional Neotropic migrant, breeding north to northern Alberta; all populations are migratory. Winters from northern Texas south to Panama, and in the Caribbean region. Common in the Central Flyway; U.S. and Canada population trend +0.8% (NS).

Canada Warbler: Temperate Nearctic/Neotropic (TG). Long-distance Neotropic migrant, breeding north to northern British Columbia; all populations are migratory. Winters in South America, from Colombia to Brazil. Common in the Central Flyway; U.S. and Canada population trend –2.2%.

Wilson's Warbler: Temperate Nearctic/Nearctic–Neotropic (CG). Long-distance transitional Neotropic migrant, Long-distance Neotropic migrant, breeding north to northern Alaska and northern Northwest Territories; all populations are migratory.

Winters from northern Mexico to western Panama. Common in the Central Flyway; U.S. and Canada population trend –2.0%.

Yellow-breasted Chat: Temperate Nearctic/Nearctic–Neotropic (TG). Short-distance transitional Neotropic migrant, breeding north to central Saskatchewan; all populations are migratory. Winters from northern Mexico to Panama. As evidence of site-fidelity in a Neotropic migrant, among 306 birds banded in western Nebraska, 1.3 percent were recaptured the following year, 0.6 percent the second year, 0.9 percent during the third year, 0.6 percent the fourth year, and 0.3 percent the sixth year (Scharf *et al*, 2008). Common in the Central Flyway; U.S. and Canada population trend –0.4%. (NB)

Family Emberizidae: Towhees & Sparrows

Spotted Towhee: Temperate Nearctic/Temperate Nearctic. Variable-distance Nearctic migrant, breeding north to central Manitoba and central Saskatchewan; the more northerly populations the most migratory; the southern populations are relatively sedentary. Winters or is residential from northern Nebraska to southern Mexico. Common in the western Central Flyway, intergrading genetically with the following species. U.S. and Canada population trend +0.2% (NS).

Eastern Towhee: Temperate Nearctic/Temperate Nearctic. Variable-distance Nearctic migrant, breeding north to southern Manitoba; the more northerly populations the most migratory; the southern populations are relatively sedentary. Winters from northern Iowa to southern Texas. Common in the eastern Central Flyway, intergrading genetically with the previous species. U.S. and Canada population trend –1.3%.

Green-tailed Towhee: Temperate Nearctic/Southern Nearctic. Variable-distance Nearctic migrant, breeding north to central Montana; the more northerly populations the most migratory; the southern populations are relatively sedentary. Winters from western Texas south to central Mexico. Common in the western Central Flyway.

Cassin's Sparrow: Temperate Nearctic/Southern Nearctic. Variable-distance Nearctic migrant, breeding north to northern Colorado; the more northerly populations the most migratory; the southern populations are relatively sedentary, although in dry years the southern birds may migrate north to breed. Winters from western Texas south to central Mexico. Common in the western Central Flyway, and a Central Flyway endemic. U.S. and Canada population trend –0.4% (NS). (NB)

Bachman's Sparrow: Southern Nearctic/Southern Nearctic. Variable-distance Nearctic migrant, breeding north to southern Missouri; the northernmost populations somewhat migratory; the southern populations are relatively sedentary. Winters from northeastern Texas south to the Gulf Coast. Local in southeastern Central Flyway; U.S. and Canada population trend –3.1%.

American Tree Sparrow: High-latitude Nearctic/Temperate Nearctic. Long-distance Nearctic migrant, breeding north to northernmost continental North America; all populations are migratory. Winters from southernmost Canada south to Oklahoma and northern Texas. Evidence from birds banded in Kansas and reported elsewhere in the Central Flyway states or internationally, or banded elsewhere but reported in Kansas (Table 1) includes movements to Kansas or from there to as far north as Alberta, Saskatchewan and Manitoba and as far south as Oklahoma. Of 35 birds banded in Oklahoma, or banded elsewhere but recovered in Oklahoma, the maximum known dispersal distance for a banded bird was from Montana (Baumgartner and Baumgartner, 1993). Records from birds banded or recovered in South Dakota range geographically from North Dakota to Oklahoma, plus Michigan (Tallman, Swanson & Palmer, 2002). Common in the Central Flyway.

Chipping Sparrow: Temperate Nearctic/Temperate Nearctic. Variable-distance Nearctic migrant, breeding north to northern Alaska and northern Northwest Territories; most populations are migratory; the southernmost populations are relatively sedentary. Winters or is resident from Oklahoma and Texas south to northern Mexico. One bird, banded in Oklahoma, was recovered in Minnesota (Baumgartner and Baumgartner, 1993). Com-

mon in the Central Flyway; U.S. and Canada population trend
–0.4%. (NB)

Clay-colored Sparrow: Temperate Nearctic/Southern Nearctic. Long-
distance Nearctic migrant, breeding north to the central Northwest Territories; all populations are migratory. Winters from
southern Texas and northern Mexico south to southern Mexico.
One bird, banded in Alberta, was recovered there later at a minimum age of six years, 11 months. Common in the Central Flyway; U.S. and Canada population trend –1.2%.

Brewer's Sparrow: Temperate Nearctic/Southern Nearctic. Variable-distance Nearctic migrant, breeding north (as an ecologi-
cally distinctive subspecies or sibling species, *S. b taverneri*) to
central Alaska; nearly all populations are migratory. Winters
from southern New Mexico south to central Mexico. One bird,
banded in Colorado, was recovered there later at a minimum
age of five years, two months. Common in the western Central
Flyway; U.S. and Canada population trend –0.1% (NS).

Field Sparrow: Temperate Nearctic/Temperate Nearctic. Variable-distance Nearctic migrant, breeding north to southern Saskatchewan. Most populations are migratory; the southernmost populations are relatively sedentary. Winters or is residential from
Kansas south to northern Mexico. Common in the Central Flyway; U.S. and Canada population trend –0.5%.

Sage Sparrow: Temperate Nearctic/Southern Nearctic. Variable-distance Nearctic migrant, breeding north to central Washington.
Most populations are migratory; the southernmost populations
are relatively sedentary. Winters from Nevada and Utah south
to northern Mexico. Common in the western Central Flyway;
U.S. and Canada population trend 0.0 (NS). (NB)

Vesper Sparrow: Temperate Nearctic/Southern Nearctic. Short-distance Nearctic migrant, breeding north to the Northwest Territories. All populations are migratory. Winters from Oklahoma and
northern Texas south to southern Mexico. Evidence from birds
banded in Kansas and reported elsewhere in the Central Flyway states or internationally, or banded elsewhere but reported

in Kansas (Table 1) includes a banded bird record of a Kansas-banded bird in Manitoba. One bird, banded in Colorado, was recovered there later at a minimum age of seven years, one month. Common in the Central Flyway; U.S. and Canada population trend –0.8%.

Lark Sparrow: Temperate Nearctic/Southern Nearctic. Variable-distance Nearctic migrant, breeding north to central Saskatchewan. Most populations are migratory; the southernmost populations are relatively sedentary. Winters from northern Texas south to southern Mexico. A bird banded in South Dakota was recovered in Michoacan, Mexico (Tallman, Swanson & Palmer, 2002). As evidence of site-fidelity in a medium-distance migrant, among 314 birds banded in western Nebraska, 5.4 percent were recaptured the following year, 2.5 percent the second year, and 0.6 percent during the third year (Scharf et al, 2008). Common in the Central Flyway; U.S. and Canada population trend –1.0%. (NB)

Lark Bunting: Temperate Nearctic/Southern Nearctic. Short-distance Nearctic migrant, breeding north to southern Saskatchewan. All populations are migratory. Winters from southern Oklahoma south to central Mexico. One bird, banded in Colorado, was recovered there later at a minimum age of four years. Common in the Central Flyway, and a Central Flyway endemic; U.S. and Canada population trend –4.4%.

Savannah Sparrow: Temperate Nearctic/Southern Nearctic. Variable-distance migrant, breeding north to northernmost continental North America. Most populations are migratory; the southernmost populations are relatively sedentary. Winters or is residential from southern Kansas south to Belize. Common in the Central Flyway; U.S. and Canada population trend –1.1%. (NB)

Grasshopper Sparrow: Temperate Nearctic/Southern Nearctic. Variable-distance Nearctic migrant, breeding north to central Saskatchewan and the southern Manitoba. Most populations are migratory; the southernmost populations are relatively sedentary. Winters or is residential from northern Texas south to northern Central America. One bird, banded in Nebraska, was recovered there later at a minimum age of three years, one month. Common in the Central Flyway; U.S. and Canada population trend –2.4%. (NB)

Baird's Sparrow: Temperate Nearctic/Southern Nearctic. Long-distance Nearctic migrant, breeding north to central Alberta and central Saskatchewan. All populations are migratory. Winters from southern Arizona south to northern Mexico. Common in the Central Flyway, and a Central Flyway endemic; U.S. and Canada population trend –2.5%.

Henslow's Sparrow: Temperate Nearctic/Southern Nearctic. Short-distance Nearctic migrant, breeding north to central Wisconsin. All populations are migratory. Winters from southern Arkansas to southern Texas. Common in the Central Flyway; U.S. and Canada population trend –0.6% (NS).

Le Conte's Sparrow: Temperate Nearctic/Southern Nearctic. Long-distance Nearctic migrant, breeding north to southern Northwest Territories. All populations are migratory. Winters from southern Kansas to southern Texas. Common in the Central Flyway; U.S. and Canada population trend –1.4% (NS).

Nelson's Sparrow: Temperate Nearctic/Southern Nearctic. Long-distance Nearctic migrant, breeding north to southern Northwest Territories. All populations are migratory. Winters along the Gulf coast of Texas and Louisiana. Uncommon in the eastern Central Flyway; U.S. and Canada population trend +1.1% (NS).

Fox Sparrow: Temperate Nearctic/Southern Nearctic. Long-distance Nearctic migrant, breeding north to northern Alaska and northern Northwest Territories. All non-coastal populations are migratory. Winters in the Great Plains from southern Kansas to southern Texas. Common in the Central Flyway; U.S. and Canada population trend +0.4% (NS).

Song Sparrow: Temperate Nearctic/Temperate Nearctic. Variable-distance Nearctic migrant, breeding north to southern Northwest Territories. Most populations are migratory; the southernmost populations are relatively sedentary. Winters or is residential from southern South Dakota to northern Texas. One bird, banded in Colorado, was recovered there later at a minimum age of 11 years, four months. Of three birds banded in Oklahoma, or banded elsewhere but recovered in Oklahoma, the maximum known dispersal distance for a banded bird was from Saskatchewan (Baumgartner and Baumgartner, 1993). A bird banded in South Dakota was recovered in Oklahoma (Tallman, Swanson & Palmer, 2002). Common in the Central Flyway; U.S. and Canada population trend +0.5%.

Lincoln's Sparrow: Temperate Nearctic/Southern Nearctic. Long-distance Nearctic migrant, breeding north to northern Alaska and northern Northwest Territories. All interior populations are migratory. Winters or is residential in the Great Plains from southern Kansas to Panama. A bird banded in South Dakota was recovered in Saskatchewan (Tallman, Swanson & Palmer, 2002). Common in the Central Flyway; U.S. and Canada population trend −1.3%.

Swamp Sparrow: Temperate Nearctic/Temperate Nearctic. Variable-distance Nearctic migrant, breeding north to central Northwest Territories. Nearly all populations are migratory; the southeasternmost populations are relatively sedentary. Winters from southern Nebraska to central Mexico. Common in the Central Flyway; U.S. and Canada population trend +0.8% (NS).

White-throated Sparrow: High-latitude Nearctic/Temperate Nearctic. Variable-distance Nearctic migrant, breeding north to central Northwest Territories. Nearly all populations are migratory; the southeasternmost populations are relatively sedentary. Winters from southern Kansas to southern Texas. One bird, banded in Alberta, was recovered there later at a minimum age of 14 years, 11 months. Of seven birds banded in Oklahoma, or banded elsewhere but recovered in Oklahoma, the maximum known dispersal distance for a banded bird was from Minnesota (Baumgartner and Baumgartner, 1993). A bird banded in South Dakota was recovered in Saskatchewan six days later (Tallman, Swanson & Palmer, 2002). Common in the Central Flyway; U.S. and Canada population trend −0.4% (NS).

Harris's Sparrow: High-latitude Nearctic/Temperate Nearctic. Long-distance Nearctic migrant, breeding north to northern Northwest

Territories. All populations are migratory. Winters in the Great
Plains from southern South Dakota to central Texas. Evidence
from birds banded in Kansas and reported elsewhere in the Central Flyway states or internationally, or banded elsewhere but
reported in Kansas (Table 1) includes movements to Kansas or
from there to as far north as Saskatchewan and as far south as
Texas. One bird, banded in Kansas, was recovered there later at
a minimum age of 11 years, eight months. Of 25 birds banded in
Oklahoma, or banded elsewhere but recovered in Oklahoma, the
maximum known dispersal distance for a banded bird was from
Saskatchewan (Baumgartner and Baumgartner, 1993). A bird
banded in South Dakota was recovered in Manitoba five days
later (Tallman, Swanson & Palmer, 2002). Common in the Central
Flyway, and a Central Flyway endemic.

White-crowned Sparrow: High-latitude Nearctic/Temperate Nearctic. Variable-distance Nearctic migrant, breeding north to
northern Alaska and northern Northwest Territories. Nearly
all interior populations are migratory. Winters from southern Nebraska to central Mexico. Of 35 birds banded elsewhere,
the maximum known dispersal distance for an Oklahoma
banding documentation was from Indiana (Baumgartner and
Baumgartner, 1993). A bird banded in South Dakota was recovered in Manitoba six days later (Tallman, Swanson & Palmer,
2002). Common in the Central Flyway; U.S. and Canada population trend −0.9%.

Dark-eyed Junco: Temperate Nearctic/Temperate Nearctic. Variable-
distance Nearctic migrant, breeding north to northern Alaska
and northern Northwest Territories. Most populations are migratory; the southernmost populations are relatively sedentary.
Winters from southern Manitoba to northern Mexico. Evidence
from birds banded in Kansas and reported elsewhere in the
Central Flyway states or internationally, or banded elsewhere
but reported in Kansas (Table 1) indicated movements to Kansas
or from there to as far north as Saskatchewan and Manitoba and
as far south as Texas. One bird, banded in Kansas, was recovered there later at a minimum age of ten years, eight months. Of
seven birds banded in Oklahoma, or banded elsewhere but re-

covered in Oklahoma, the maximum known dispersal distance for a banded bird was from Saskatchewan (Baumgartner and Baumgartner, 1993). Records from birds banded or recovered in South Dakota range geographically from Alaska and Alberta to Oklahoma, and east to New York (Tallman, Swanson & Palmer, 2002). Common in the Central Flyway; U.S. and Canada population trend –1.2%. (NB)

Family Cardinalidae: Cardinals, Tanagers & Grosbeaks

Summer Tanager: Temperate Nearctic/Neotropic (TG). Short-distance to long-distance Neotropic migrant, breeding north to southern Iowa. All populations are migratory. Winters or is residential from southernmost Texas to Ecuador and Brazil, and in the Caribbean region. One bird, banded in Texas, was recovered there later at a minimum age of seven years, three months. Common in the southern Central Flyway; U.S. and Canada population trend +0.1% (NS). (NB)

Scarlet Tanager: Temperate Nearctic/Neotropic (TG). Long-distance Neotropic migrant, breeding north to southern Manitoba. All populations are migratory. Winters in northern South America, from Colombia to northern Bolivia. Common in the eastern Central Flyway; U.S. and Canada population trend –0.1% (NS).

Western Tanager: Temperate Nearctic/Northern Neotropic. Short-distance to long-distance Neotropic migrant, breeding north to central Northwest Territories. All populations are migratory. Winters from northern Mexico to western Panama. Common in the western Central Flyway; U.S. and Canada population trend +1.2%. (NB). Figure 47.

Hepatic Tanager: Temperate Nearctic/Neotropic. Short-distance to long-distance Neotropic migrant, breeding north to southeastern Colorado. All populations are migratory. Winters or is residential from northern Mexico to Ecuador and Argentina. Uncommon to rare in the southwestern Central Flyway; U.S. and Canada population trend +1.7% (NS).

Figure 47. Western Tanager, adult male.

Rose-breasted Grosbeak: Temperate Nearctic/Northern Neotropic
(TG). Long-distance Neotropic migrant, breeding north to central Northwest Territories. All populations are migratory. Winters or is residential from northern Mexico to Peru. Common in
the eastern Central Flyway; U.S. and Canada population trend
–0.6%.

Black-headed Grosbeak: Temperate Nearctic/Nearctic–Neotropic. Variable-distance transitional Neotropic migrant, breeding north to central British Columbia. All U.S. populations are migratory. Winters from northern Mexico to southern Mexico. One bird, banded in Montana, was recovered there later at a minimum age of 11 years, 11 months. Common in the western Central Flyway; U.S. and Canada population trend +1.0%. (NB)

Blue Grosbeak: Temperate Nearctic/Northern Neotropic (TG). Variable-distance Neotropic migrant, breeding north to central North Dakota. All U.S. populations are migratory. Winters from northwestern Mexico to Costa Rica. Common in the Central Flyway; U.S. and Canada population trend +1.0%. (NB)

Lazuli Bunting: Temperate Nearctic/Nearctic-Neotropic. Short-distance transitional Neotropic migrant, breeding north to southern British Columbia. All U.S. populations are migratory. Winters from northwestern Mexico to central Mexico. Common in the western Central Flyway; U.S. and Canada population trend –0.2% (NS). (NB)

Indigo Bunting: Temperate Nearctic/Nearctic-Neotropic (TG/CG). Short-distance to long-distance transitional Neotropic migrant, breeding north to southern Manitoba. All U.S. populations are migratory. Winters from southern Texas to Panama, and in the Caribbean region. As evidence of site-fidelity in a Neotropic migrant, among 44 birds banded in western Nebraska, 6.8 percent were recaptured the following year (Scharf *et al*, 2008). Common in the Central Flyway; U.S. and Canada population trend +1.5%.

Varied Bunting: Southern Nearctic/Nearctic-Neotropic. Variable-distance transitional Neotropic migrant, breeding north to southern Texas. All U.S. populations are migratory. Winters from northwestern Mexico to southern Mexico. Local in the southern Central Flyway.

Painted Bunting: Southern Nearctic/Nearctic-Neotropic (TG/CG). Short-distance to long-distance transitional Neotropic migrant, breeding north to southern Kansas. All U.S. populations are migratory. Winters from northeastern Mexico to central Panama, and in the Caribbean region. One bird, banded in Texas, was recovered there later at a minimum age of 11 years. Common in the Central Flyway; U.S. and Canada population trend –0.1% (NS). (NB)

Dickcissel: Temperate Nearctic/Nearctic-Neotropic (TG). Long-distance Neotropic migrant, breeding north to northern North Dakota. All populations are migratory. Winters from western Mexico (erratically) to northern South America, from Colombia and Venezuela east to Suriname. One bird, banded in Kansas, was recovered there later at a minimum age of four years. Common in the Central Flyway; U.S. and Canada population trend –0.5% (NS).

Family Icteridae: Blackbirds & Orioles

Bobolink: Temperate Nearctic/Southern Neotropic (TG). Long-distance Neotropic migrant, breeding north to northern Alberta. All populations are migratory. Winters in southern South America, mainly in Argentine grasslands. Common in the Central Flyway; U.S. and Canada population trend +2.2%.

Red-winged Blackbird: Temperate Nearctic/Temperate Nearctic. Variable-distance Nearctic migrant, breeding north to central Alaska and central Northwest Territories. Northern populations are migratory; the southernmost populations are relatively sedentary. Winters or is residential from southern North Dakota to Cost Rica. Evidence from birds banded in Kansas and reported elsewhere in the Central Flyway states or internationally, or banded elsewhere but reported in Kansas (Table 1) includes movements to Kansas or from there to as far north as Alberta, Saskatchewan and Manitoba, and as far south as Texas. Of 79 birds banded in Oklahoma, or banded elsewhere but recovered in Oklahoma, the maximum known dispersal distance for a banded bird was from Alberta (Baumgartner and Baumgartner, 1993). Of 120 interstate recoveries from birds banded in South Dakota, 30 percent were from North Dakota and 23 percent from Texas, but recoveries ranged north as

Figure 48. Red-winged Blackbird, adult male.

far as Alberta (Tallman, Swanson & Palmer, 2002). As evidence of site-fidelity in a short-distance Nearctic migrant, among 1,074 birds banded in western Nebraska, 1.0 percent were recaptured the following year, 2.1 percent the second year, 0.4 percent the third year, and 0.1 percent the fourth year (Scharf *et al*, 2008). Common in the Central Flyway; U.S. and Canada population trend –0.9%. Figure 48.

Eastern Meadowlark: Temperate Nearctic/Temperate Nearctic. Variable-distance Nearctic migrant, breeding north to northern
Minnesota. Northern populations are migratory; the southern populations are relatively sedentary. Winters or is residential from Kansas to Mexico, locally to South America. Of seven
birds banded in Oklahoma, or banded elsewhere but recovered in Oklahoma, the maximum known dispersal distance for a
banded bird was from Minnesota (Baumgartner and Baumgartner, 1993). Common in the Central Flyway; U.S. and Canada
population trend +1.5%.

Western Meadowlark: Temperate Nearctic/Temperate Nearctic. Variable-distance Nearctic migrant, breeding north to central Alberta. Northern populations are migratory; the southern populations are relatively sedentary. Winters or is residential from
Nebraska to central Mexico. Evidence from birds banded in
Kansas and reported elsewhere in the Central Flyway states
or internationally, or banded elsewhere but reported in Kansas (Table 1) includes the movement of a Kansas-banded bird to
Texas, One bird, banded in Colorado, was recovered there later
at a minimum age of six years, six months. Of two birds banded
in Oklahoma, or banded elsewhere but recovered in Oklahoma,
the maximum known dispersal distance for a banded bird was
from Saskatchewan (Baumgartner and Baumgartner, 1993).
Common in the Central Flyway; U.S. and Canada population
trend –1.0%. (NB)

Yellow-headed Blackbird: Temperate Nearctic/Southern Nearctic.
Variable-distance Nearctic migrant, breeding north to northern Alberta. Nearly all populations are migratory. Winters from
New Mexico to central Mexico. Evidence from birds banded
in Kansas and reported elsewhere in the Central Flyway states
or internationally, or banded elsewhere but reported in Kansas (Table 1) includes movements to Kansas or from there to
as far north as North Dakota and as far south as Mexico. One
bird, banded in Saskatchewan, was recovered later in Nebraska at a minimum age of ten years, 11 months. Of two birds
banded elsewhere, the maximum known dispersal distance
for an Oklahoma banding documentation was from South Da-

Figure 49. Yellow-headed Blackbird, adult male.

kota (Baumgartner and Baumgartner, 1993). Records from birds banded or recovered in South Dakota range geographically from Saskatchewan to Texas, and west to Arizona (Tallman, Swanson & Palmer, 2002). Common in the Central Flyway; U.S. and Canada population trend –0.4% (NS). (NB), Figure 49.

Rusty Blackbird: Temperate Nearctic/Temperate Nearctic. Short-distance Nearctic migrant, breeding north to northern Alaska and northernmost Northwest Territories. All populations are migratory. Winters from Nebraska south to the Gulf Coast. Of 13 birds banded elsewhere, the maximum known dispersal distance for an Oklahoma banding documentation was from Arkansas (Baumgartner and Baumgartner, 1993). Records from birds banded or recovered in South Dakota range geographically from Saskatchewan to Arkansas, (Tallman, Swanson & Palmer, 2002). Common in the Central Flyway; U.S. and Canada population trend –3.6%.

Brewer's Blackbird: Temperate Nearctic/Temperate Nearctic. Variable-distance Nearctic migrant, breeding north to northern Alaska and northern Northwest Territories. Northern populations are migratory; the southern populations are relatively sedentary. Winters from Nebraska south to central Mexico. Common in the Central Flyway; U.S. and Canada population trend –2.1%. (NB)

Common Grackle: Temperate Nearctic/Temperate Nearctic. Variable-distance Nearctic migrant, breeding north to southern Northwest Territories. Northern populations are migratory; the southern populations are relatively sedentary. Winters from South Dakota south to Gulf Coast. Evidence from birds banded in Kansas and reported elsewhere in the Central Flyway states or internationally, or banded elsewhere but reported in Kansas (Table 1) includes movements to Kansas or from there to as far north as Saskatchewan and Ontario and as far south as Texas. One bird, banded in Minnesota, was recovered there later at a minimum age of 22 years, 11 months! Of 289 birds banded in Oklahoma, or banded elsewhere but recovered in Oklahoma, the maximum known dispersal distance for a banded bird was from Montana (Baumgartner and Baumgartner, 1993). A total of 1,102 records from birds banded or recovered in South Dakota range geographically from Montana and North Dakota to Texas, with scattered records east to Virginia and Maryland (Tallman, Swanson & Palmer, 2002). Common in the Central Flyway; U.S. and Canada population trend +1.6%.

Great-tailed Grackle: Southern Nearctic/Southern Nearctic. Variable-distance Nearctic migrant, breeding north to southern South Dakota. Northern populations are migratory; the southern populations are relatively sedentary. Winters or is residential from Kansas south to Peru. Of 10 birds banded in Oklahoma, or banded elsewhere but recovered in Oklahoma, the maximum known dispersal distance for a banded bird was from Texas (Baumgartner and Baumgartner, 1993). Common in the southern Central Flyway; U.S. and Canada population trend +2.7%.

Bronzed Cowbird: Southern Nearctic/Southern Nearctic. Variable-distance Nearctic migrant, breeding north to central Texas. Northern populations are migratory; the southern populations are relatively sedentary. Winters or is residential from southern Texas south to central Panama. Common in the southern Central Flyway; U.S. and Canada population trend +0.3% (NS).

Brown-headed Cowbird: Temperate Nearctic/Temperate Nearctic. Variable-distance Nearctic migrant, breeding north to central Northwest Territories. Northern populations are migratory; the southern populations are relatively sedentary. Winters or is residential from Kansas south to southern Mexico. Evidence from birds banded in Kansas and reported elsewhere in the Central Flyway states or internationally, or banded elsewhere but reported in Kansas (Table 1) includes movements to Kansas or from there to as far north as Saskatchewan and as far south as Mexico. Of 69 birds banded in Oklahoma, or banded elsewhere but recovered in Oklahoma, the maximum known dispersal distance for a banded bird was from Michigan (Baumgartner and Baumgartner, 1993). Records from birds banded or recovered in South Dakota range geographically from Alberta and Saskatchewan to Texas, and east to Minnesota (Tallman, Swanson & Palmer, 2002). As evidence of site-fidelity in a short-distance Nearctic migrant, among 165 birds banded in western Nebraska, 18.3 percent were recaptured the following year, and 1.2 percent the second year (Scharf *et al*, 2008). This is a remarkably high first-year recapture rates, and might reflect the importance of Cowbird females remembering and returning to familiar nest-rich habitats. Common in the Central Flyway; U.S. and Canada population trend –0.6%. (NB)

Orchard Oriole: Temperate Nearctic/Nearctic-Neotropic (TG). Short- to long-distance transitional Neotropic migrant, breeding north to southern Saskatchewan. All U.S. populations are migratory. Winters or resident from northern Mexico to northern South America. Common in the Central Flyway; U.S. and Canada population trend –0.9%. One bird, banded in Nebraska, was recovered there later at a minimum age of ten years, 11 months. As evidence of site-fidelity in a Neotropic migrant, among 1,829 birds banded in western Nebraska, 1.5 percent were recaptured the following year, 0.9 percent the second year, 0.4 percent the third year, 0.1 percent the fourth year, and 0.05 percent the sixth year (Scharf *et al*, 2008). (NB)

Bullock's Oriole: Temperate Nearctic/Nearctic Neotropic. Short- to long-distance transitional Neotropic migrant, breeding north to central British Columbia. All U.S. populations are migratory. Winters or resident from northern Mexico to Guatemala. One bird, banded in Colorado, was recovered there later at a minimum age of eight years, 11 months. Common in the western Central Flyway; U.S. and Canada population trend –0.4% (NS). (NB)

Baltimore Oriole: Temperate Nearctic/Nearctic-Neotropic (TG). Long-distance transitional Neotropic migrant, breeding north to northern British Columbia. All U.S. populations are migratory. Winters from northern Mexico to Colombia and Venezuela. One bird, banded in Minnesota, was recovered there later at a minimum age of 12 years. One banded in South Dakota was recovered two month later in Costa Rica (Tallman, Swanson & Palmer, 2002). Records from birds banded or recovered in South Dakota range geographically from North Dakota to Nebraska, and east to Quebec (Tallman, Swanson & Palmer, 2002). As evidence of site-fidelity in a Neotropic migrant, among 284 birds banded in western Nebraska, 1.8 percent were recaptured the following year, and 0.1 percent the second year (Scharf *et al*, 2008). Common in the eastern Central Flyway; U.S. and Canada population trend –1.2%. (NB)

Hooded Oriole: Southern Nearctic/Nearctic–Neotropic. Variable-distance transitional Neotropic migrant, breeding north to west-central Texas. All U.S. populations are migratory. Winters or is residential from northern Mexico to Belize. Common in the southernmost Central Flyway; U.S. and Canada population trend +0.3% (NS).

Scott's Oriole: Southern Nearctic/Nearctic–Neotropic. Variable-distance transitional Neotropic migrant, breeding north to southern Wyoming. All U.S. populations are migratory. Winters or is residential from northern Mexico to southern Mexico. Common in the southernmost Central Flyway; U.S. and Canada population trend –0.4% (NS).

Family Fringillidae: Finches

Gray-crowned Rosy-Finch: Extralimital/Altitudinal. Variable-distance Nearctic migrant, breeding north to northern Alaska and northern Northwest Territories. All populations are latitudinally or altitudinally migratory. Winters or is residential from British Columbia to New Mexico, sometimes nomadic. Two birds banded in South Dakota were later recovered in Wyoming (Tallman, Swanson & Palmer, 2002). Occasional (as nomadic vagrants) in the western Central Flyway.

Pine Grosbeak: Temperate Nearctic/Temperate Nearctic. Irruptive Holarctic migrant; residential north to northern Alaska and northern Northwest Territories. Irruptive and nomadic, as the birds are largely dependent on pine seeds. Because of this food dependence, few Pine Grosbeaks occur in the central Flyway. Occasional in the western Central Flyway; U.S. and Canada population trend –1.7%.

Purple Finch: Temperate Nearctic/Temperate Nearctic Variable-distance and irruptive Nearctic migrant, breeding north to central Northwest Territories. Northern populations are migratory and irruptive, as the birds are largely dependent on tree seeds, wintering variably far south, sometimes almost to Gulf Coast. Evidence from birds banded in Kansas and reported elsewhere in the Central Flyway states or internationally, or banded elsewhere but reported in Kansas (Table 1) includes indicated movements to Kansas or from there to as far north as Alberta, Saskatchewan and Manitoba, and as far south as Oklahoma. Of 77 birds banded in Oklahoma, or banded elsewhere but recovered in Oklahoma, the maximum known dispersal distance for a banded bird was from Maine (Baumgartner and Baumgartner, 1993). Records from birds banded or recovered in South Dakota range geographically from Kansas east to Iowa and Ontario (Tallman, Swanson & Palmer, 2002). Common in the northern Central Flyway; rarer or more periodic southwardly. U.S. and Canada population trend –1.4%.

Cassin's Finch: Temperate Nearctic/Temperate Nearctic. Variable-distance Nearctic migrant, breeding north to central British Columbia. Northern populations are somewhat migratory, wintering variably south, sometimes to southern Arizona and New Mexico. Occasional to rare in the western Central Flyway; U.S. and Canada population trend –2.5%. (NB)

House Finch: Temperate Nearctic/Temperate Nearctic. Relative sedentary Nearctic migrant, increasingly migratory at northern parts of range. Evidence from birds banded in Kansas and reported elsewhere in the Central Flyway states or internationally, or banded elsewhere but reported in Kansas (Table 1) includes indicated movements to Kansas or from there to as far north as North Dakota and as far south as Texas. Records from birds banded or recovered in South Dakota range geographically from North Dakota to Kansas, and east to Wisconsin (Tallman, Swanson & Palmer, 2002). As evidence of apparent very low site-fidelity in a semi-residential species, of 1,302 birds banded in western Nebraska, 0.009 percent were recaptured the following year, 0.003 percent the second year, and 0.001 percent the third year (Scharf et al, 2008). This very low fidelity rate might have resulted from the species being a rapidly increasing and geographically expanding population in western Nebraska during the early 2000's. Common in the Central Flyway; U.S. and Canada population trend +0.3% (NS).

Red Crossbill: Temperate Nearctic/Temperate Nearctic. Irruptive Holarctic migrant; residential north to northern Alaska and northern Northwest Territories. Sometimes irruptive and nomadic, often reaching the southern Great Plains. In North America as many as eight different genetic Red Crossbill types are present, each distinctive both structurally as to its beak shape and behaviorally in terms of its preferred seeds. Each crossbill type is adapted to a particular coniferous food source, and all tend to occupy different areas, although all types can exploit several species of conifers (Benkman, 1990). Because of their dependence on conifer seeds, crossbill migrations are dependent of available coniferous food sources (Newton, 2008). As a result they are only occasionally found in the Central Flyway, mostly in the north. U.S. and Canada population trend –0.5% (NS).

White-winged Crossbill: Temperate Nearctic/Temperate Nearctic. Irruptive Holarctic migrant; residential north to southern Alaska and the southern Northwest Territories. Sometimes irruptive and nomadic, occasionally reaching the central Great Plains. Movements of this crossbill species are largely associated with seed supplies of tamarack and spruces, northern-oriented conifers whose cones are relatively easy to handle, while the highly diverse beak shapes of Red Crossbills allow them to pry open a greater variety of conifer cones, including even those of many and widely distributed species of pines (Benkman, 1990, Newton, 2008). As a result, White-winged Crossbills are rare in the southern Central Flyway, but are progressively more frequent northwardly. U.S. and Canada population trend +1.3% (NS).

Common Redpoll: High-latitude Nearctic/Temperate Nearctic. Variable-distance Holarctic migrant, breeding north to northern Alaska and northern Northwest Territories. Northern populations are migratory and irruptive, as the birds are largely dependent on birch and alder seeds. The Hoary Redpoll is an arctic-breeding species that rarely reaches the Great Plains, but it too is irruptive, and is sometimes found in company with the Common Redpoll. The latter winters from central Canada variably south to northern U. S, sometimes reaching the northern Great Plains. One bird, banded in Manitoba, was recovered later in Wisconsin at a minimum age of five years, nine months. Records from birds banded or recovered in South Dakota range geographically from Wisconsin east to Ontario, Quebec and Massachusetts (Tallman, Swanson & Palmer, 2002), with no evidence of concentration in the Central Flyway. Uncommon in the northern Central Flyway.

Pine Siskin: Temperate Nearctic/Temperate Nearctic. Variable-distance Holarctic migrant, breeding north to southern Alaska and the southern Northwest Territories. Northern populations are migratory and irruptive, as the birds are largely dependent on pine, birch and alder seeds, wintering from central Canada variably south to southern Great Plains. Evidence from birds banded in Kansas and reported elsewhere in the Central Flyway states or internationally, or banded elsewhere but reported in Kansas (Table 1) includes indicated movements to Kansas or from there to as far north as Alberta, Saskatchewan and Manitoba and as far south as Texas. Of five birds banded in Oklahoma, or banded elsewhere but recovered in Oklahoma, the maximum known dispersal distance for a banded bird was from Michigan (Baumgartner and Baumgartner, 1993). Records from birds banded or recovered in South Dakota range geographically from British Columbia and Alberta south to Kansas, with scattered records east to Connecticut and west to California (Tallman, Swanson & Palmer, 2002). Common in the central and northern Central Flyway; U.S. and Canada population trend –2.9%. (NB)

Lesser Goldfinch: Temperate Nearctic/Southern Nearctic. Variable-distance Nearctic migrant, breeding irregularly north to eastern Wyoming and southern South Dakota. Northern populations are migratory, wintering from central Canada variably south to northeastern Mexico. Common in the southern Central Flyway; U.S. and Canada population trend +1.0%.

American Goldfinch: Temperate Nearctic/Temperate Nearctic. Variable-distance Nearctic migrant, breeding north to central Alberta and Saskatchewan. Northern populations are migratory, wintering from central Canada variably south to northeastern Mexico. Evidence from birds banded in Kansas and reported elsewhere in the Central Flyway states or internationally, or banded elsewhere but reported in Kansas (Table 1) includes movements to Kansas or from there to as far north as Saskatchewan and as far south as Texas. One bird, banded in Missouri, was recovered later in Wisconsin, at a minimum age of nine years, two months. Of 19 birds banded in Oklahoma, or banded elsewhere but recovered in Oklahoma, the maximum known dispersal distance for a banded bird was from Manitoba (Baumgartner and Baumgartner, 1993). Records from birds banded or recovered in South Dakota range geographically from Saskatchewan to Texas, and east to Wisconsin (Tallman, Swanson & Palmer, 2002). As evidence of site-fidelity in a seemingly residential species, of 950 birds banded in western Nebraska, 3.5 percent were recaptured the following year, 1.6 percent the second year, and 0.001 percent the third year, (Scharf *et al,* 2008). Common in the Central Flyway; U.S. and Canada population trend stable: 0.0%.

Evening Grosbeak: Temperate Nearctic/Temperate Nearctic. Irruptive Nearctic migrant; breeding north to southern Northwest Territories. Mostly residential, but winters regularly from southern Canada to the northern Great Plains, from Montana to Minnesota, Sometimes irruptive and nomadic, occasionally reaching the southern Great Plains. Of five birds banded in Oklahoma, or banded elsewhere but recovered in Oklahoma, the maximum known dispersal distance for a banded bird was from New York (Baumgartner and Baumgartner, 1993). Records from birds banded or recovered in South Dakota range geographically from Washington and Oregon east to New Jersey (Tallman, Swanson & Palmer, 2002). This species is in serious national decline, and major population irruptions reaching the Great Plains no longer occur. According to Newton (2008), this species is part of a group of boreal seed-eating species that tend to irrupt together, and also include the. Common Redpoll, Pine Siskin, Purple Finch, Red-breasted Nuthatch, Black-capped Chickadee, Red Crossbill and White-winged Crossbill. Occasional in the Central Flyway, mostly in the north. U.S. and Canada population trend –1.8%.

Appendix.

Taxonomic List of Species Mentioned in the Text

Family Anatidae: Swans, Geese, and Ducks (33 migratory spp.)
  Fulvous Whistling-Duck, *Dendrocygna bicolor*
  Black-bellied Whistling-Duck, *Dendrocygna autumnalis*
  Greater White-fronted Goose, *Anser albifrons*
  Snow Goose, *Chen caerulescens*
  Ross's Goose, *Chen rossii*
  Cackling Goose, *Branta hutchinsii*
  Canada Goose, *Branta canadensis*
  Trumpeter Swan, *Cygnus buccinator*
  Tundra Swan, *Cygnus columbianus*
  Wood Duck, *Aix sponsa*
  Gadwall, *Anas strepera*
  American Wigeon, *Anas americana*
  American Black Duck, *Anas rubripes*
  Mallard, *Anas platyrhynchos*
  Mottled Duck, *Anas fulvigula*
  Blue-winged Teal, *Anas discors*
  Cinnamon Teal, *Anas cyanoptera*
  Northern Shoveler, *Anas clypeata*
  Northern Pintail, *Anas acuta*
  Green-winged Teal, *Anas crecca*
  Canvasback, *Aythya valisineria*
  Redhead, *Aythya americana*
  Ring-necked Duck, *Aythya collaris*
  Greater Scaup, *Aythya marila*
  Lesser Scaup, *Aythya affinis*
  Surf Scoter, *Melanitta perspicillata*
  White-winged Scoter, *Melanitta fusca*
  Long-tailed Duck, *Clangula hyemalis*
  Bufflehead, *Bucephala albeola*
  Common Goldeneye, *Bucephala clangula*
  Hooded Merganser, *Lophodytes cucullatus*

Common Merganser, *Mergus merganser*
Red-breasted Merganser, *Mergus serrator*
Ruddy Duck, *Oxyura jamaicensis*

Family Gaviidae: Loons (3 migratory spp.)
Red-throated Loon, *Gavia stellata*
Pacific Loon, *Gavia pacifica*
Common Loon, *Gavia immer*
Yellow-billed *Loon, Gavia adamsi*

Family Podicipedidae: Grebes (5 migratory spp.)
Pied-billed Grebe, *Podilymbus podiceps*
Horned Grebe, *Podiceps auritus*
Red-necked Grebe, *Podiceps grisigena*
Eared Grebe, *Podiceps nigricollis*
Western Grebe, *Aechmophorus occidentalis*
Clark's Grebe, *Aechmophorus clarkii*

Family Phalacrocoracidae: Cormorants (1 migratory sp.)
Double-crested Cormorant, *Phalacrocorax auritus*

Family Anhingidae: Anhingas (1 migratory sp.)
Anhinga: *Anhinga anhinga*

Family Pelecanidae: Pelicans (1 migratory sp.)
American White Pelican, *Pelecanus erythrorhynchos*

Family Ardeidae: Bitterns and Herons (12 migratory spp.)
American Bittern, *Botaurus lentiginosus*
Least Bittern, *Ixobrychus exilis*
Great Blue Heron, *Ardea herodias*
Great Egret, *Ardea alba*
Snowy Egret, *Egretta thula*
Little Blue Heron, *Egretta caerulea*
Tricolored Heron, *Egretta tricolor*
Reddish Egret, *Egretta rufescens*
Cattle Egret, *Bubulcus ibis*
Green Heron, *Butorides virescens*
Black-crowned Night-Heron, *Nycticorax nycticorax*
Yellow-crowned Night-Heron, *Nyctanassa violacea*

Family Threskiornithidae: Ibises and Spoonbills (4 migratory spp.)
  White Ibis, *Eudomimus albus*
  White-faced Ibis, *Plegadis chihi*
  Glossy Ibis, *Plegadis falcinellis*
  Roseate Spoonbill, *Platalea ajaja*

Family Cathartidae: New World Vultures (1 migratory sp.)
  Turkey Vulture, *Cathartes aura*

Family Pandionidae: Ospreys (1 migratory sp.)
  Osprey, *Pandion haliaetus*

Family Accipitridae: Kites, Hawks and Eagles (17 migratory spp.)
  Swallow-tailed Kite, *Elanoides forficatus*
  White-tailed Kite, *Elanus leucurus*
  Mississippi Kite, *Ictinia mississippiensis*
  Bald Eagle, *Haliaeetus leucocephalus*
  Northern Harrier, *Circus cyaneus*
  Sharp-shinned Hawk, *Accipiter striatus*
  Cooper's Hawk, *Accipiter cooperii*
  Northern Goshawk, *Accipiter gentilis*
  Common Black-Hawk, *Buteogallus anthracinus*
  Red-shouldered Hawk, *Buteo lineatus*
  Broad-winged Hawk, *Buteo platypterus*
  Swainson's Hawk, *Buteo swainsoni*
  Zone-tailed Hawk, *Buteo albonotatus*
  Red-tailed Hawk, *Buteo jamaicensis*
  Ferruginous Hawk, *Buteo regalis*
  Rough-legged Hawk, *Buteo lagopus*
  Golden Eagle, *Aquila chrysaetos*

Family Falconidae: Falcons (5 migratory spp.)
  American Kestrel, *Falco sparverius*
  Merlin, *Falco columbarius*
  Gyrfalcon, *Falco rusticus*
  Peregrine Falcon, *Falco peregrinus*
  Prairie Falcon, *Falco mexicanus*

Family Rallidae: Rails, Gallinules and Coots (8 migratory spp.)
  Yellow Rail, *Coturnicops noveboracensis*
  Black Rail, *Laterallus jamaicensis*
  King Rail, *Rallus elegans*

Virginia Rail, *Rallus limicola*
Sora, *Porzana carolina*
Purple Gallinule, *Porphyrula martinica*
Common Gallinule, *Gallinula galeata*
American Coot, *Fulica americana*

Family Gruidae: Cranes (2 migratory spp.)
Sandhill Crane, *Grus canadensis*
Whooping Crane, *Grus americana*

Family Charadriidae: Plovers (8 migratory spp.)
Black-bellied Plover, *Pluvialis squatarola*
American Golden-Plover, *Pluvialis dominica*
Snowy Plover, *Charadrius nivosus*
Wilson's Plover, *Charadrius wilsonia*
Mountain Plover, *Charadrius montana*
Semipalmated Plover, *Charadrius semipalmatus*
Piping Plover, *Charadrius melodus*
Killdeer, *Charadrius vociferus*

Family Recurvirostridae: Stilts and Avocets (2 migratory spp.)
Black-necked Stilt, *Himantopus mexicanus*
American Avocet, *Recurvirostra americana*

Family Scolopacidae: Sandpipers and Phalaropes (29 migratory spp.)
Spotted Sandpiper, *Tringa macularia*
Solitary Sandpiper, *Tringa solitaria*
Greater Yellowlegs, *Tringa melanoleuca*
Willet, *Tringa semipalmata*
Lesser Yellowlegs, *Tringa flavipes*
Upland Sandpiper, *Bartramia longicauda*
Whimbrel, *Numenius phaeopus*
Long-billed Curlew, *Numenius americanus*
Eskimo Curlew, *Numenius borealis*
Hudsonian Godwit, *Limosa haemastica*
Marbled Godwit, *Limosa fedoa*
Ruddy Turnstone, *Arenaria interpres*
Red Knot, *Calidris canutus*
Sanderling, *Calidris alba*
Semipalmated Sandpiper, *Calidris pusilla*
Western Sandpiper, *Calidris mauri*
Least Sandpiper, *Calidris minutilla*

White-rumped Sandpiper, *Calidris fuscicollis*
Baird's Sandpiper, *Calidris bairdii*
Pectoral Sandpiper, *Calidris melanotos*
Dunlin, *Calidris alpina*
Stilt Sandpiper, *Calidris himantopus*
Buff-breasted Sandpiper, *Tryngites subruficollis*
Short-billed Dowitcher, *Limnodromus griseus*
Long-billed Dowitcher, *Limnodromus scolopaceus*
Wilson's Snipe, *Gallinago delicata*
American Woodcock, *Scolopax minor*
Wilson's Phalarope, *Phalaropus tricolor*
Red-necked Phalarope, *Phalaropus lobatus*
Red Phalarope, *Phalaropus fulicaria*

Family Laridae: Gulls and Terns (17 migratory spp.)
Bonaparte's Gull, *Coriocephalus philadelphia*
Little Gull, *Hydrocoleus minutus*
Laughing Gull, *Leucophaeus atricilla*
Franklin's Gull, *Leucophaeus pipixcan*
Mew Gull: *Larus canus*
Ring-billed Gull, *Larus delawarensis*
California Gull, *Larus californicus*
Herring Gull, *Larus argentatus*
Lesser Black-backed Gull, *Larus fuscus*
Least Tern, *Sterna antillarum*
Gull-billed Tern, *Sterna nilotica*
Caspian Tern, *Sterna caspia*
Royal Tern, *Sterna maxima*
Sandwich Tern, *Sterna sandvicensis*
Black Tern, *Chlidonias niger*
Common Tern, *Sterna hirundo*
Forster's Tern, *Sterna forsteri*

Family Columbidae: Pigeons and Doves (2 migratory spp.)
White-winged Dove, *Zenaida asiatica*
Mourning Dove, *Zenaida macroura*

Family Cuculidae: Cuckoos (2 migratory spp.)
Yellow-billed Cuckoo, *Coccyzus americanus*
Black-billed Cuckoo, *Coccyzus erythropthalmus*

Family Tytonidae: Barn Owl (1 migratory sp.)
Barn Owl, *Tyto alba*

Family Strigidae: Typical Owls (8 migratory or semi-migratory spp.)
    Great Horned Owl, *Bubo virginianus*
    Snowy Owl, *Bubo scandiaca*
    Burrowing Owl, *Athene cunicularia*
    Great Gray Owl, *Strix nebulosa*
    Long-eared Owl, *Asio otus*
    Short-eared Owl, *Asio flammeus*
    Northern Saw-whet Owl, *Aegolius acadicus*
    Boreal Owl, *Aegolius funereus*

Family Caprimulgidae: Goatsuckers (5 migratory spp.)
    Lesser Nighthawk, *Chordeiles acutipennis*
    Common Nighthawk, *Chordeiles minor*
    Common Pauraque, *Nyctidomus albicollis*
    Common Poorwill, *Phalaenoptilus nuttallii*
    Chuck-will's-widow, *Caprimulgus carolinensis*
    Eastern Whip-poor-will, *Caprimulgus vociferus*

Family Apodidae: Swifts (2 migratory spp.)
    Chimney Swift, *Chaetura pelagica*
    White-throated Swift, *Aeronautes saxatalis*

Family Trochilidae: Hummingbirds (9 migratory spp.)
    Ruby-throated Hummingbird, *Archilochus colubris*
    Black-chinned Hummingbird, *Archilochus alexandri*
    Calliope Hummingbird, *Selasphorus calliope*
    Broad-tailed Hummingbird, *Selasphorus platycercus*
    Rufous Hummingbird, *Selasphorus rufus*
    Magnificent Hummingbird, *Eugenes fulgens*
    Blue-throated Hummingbird, *Lampornis clemenciae*
    Lucifer Hummingbird, *Calothorax lucifer*
    Buff-bellied Hummingbird, *Amazilia yucatanensis*

Family Alcedinidae: Kingfishers (1 sp.)
    Belted Kingfisher, *Ceryle alcyon*

Family Picidae: Woodpeckers (6 migratory spp.)
    Lewis's Woodpecker, *Melanerpes lewis*
    Red-headed Woodpecker, *Melanerpes erythrocephalus*
    Yellow-bellied Sapsucker, *Sphyrapicus varius*
    Red-naped Sapsucker, *Sphyrapicus nuchalis*
    Williamson's Sapsucker, *Sphyrapicus thyroides*
    Northern Flicker, *Colaptes auratus*

Family Tyrannidae: American Flycatchers (24 migratory spp.)
  Northern Beardless-Tyrannulet, *Camptostoma imberbe*
  Olive-sided Flycatcher, *Contopus cooperi*
  Western Wood-Pewee, *Contopus sordidulus*
  Eastern Wood-Pewee, *Contopus virens*
  Yellow-bellied Flycatcher, *Empidonax flaviventris*
  Acadian Flycatcher, *Empidonax virescens*
  Alder Flycatcher, *Empidonax alnorum*
  Willow Flycatcher, *Empidonax traillii*
  Least Flycatcher, *Empidonax minimus*
  Black Phoebe, *Sayornis nigricans*
  Eastern Phoebe, *Sayornis phoebe*
  Say's Phoebe, *Sayornis saya*
  Vermilion Flycatcher, *Pyrocephalus rubinus*
  Ash-throated Flycatcher, *Myiarchus cinerascens*
  Great Crested Flycatcher, *Myiarchus crinitus*
  Brown-crested Flycatcher, *Myiarchus tyrannulus*
  Great Kiskadee, *Pitangus sulphuratus*
  Tropical Kingbird, *Tyrannus melancholicus*
  Couch's Kingbird, *Tyrannus couchii*
  Cassin's Kingbird, *Tyrannus vociferans*
  Western Kingbird, *Tyrannus verticalis*
  Eastern Kingbird, *Tyrannus tyrannus*
  Scissor-tailed Flycatcher, *Tyrannus forficatus*
  Rose-throated Becard, *Pachyramphus agiaiae*

Family Laniidae: Shrikes (2 migratory spp.)
  Loggerhead Shrike, *Lanius ludovicianus*
  Northern Shrike, *Lanius excubitor*

Family Vireonidae: Vireos (10 migratory spp.)
  Bell's Vireo, *Vireo bellii*
  Black-capped Vireo, *Vireo atricapillus*
  Yellow-throated Vireo, *Vireo flavifrons*
  Plumbeous Vireo, *Vireo plumbeus*
  Cassin's Vireo, *Vireo cassinii*
  Blue-headed Vireo, *Vireo solitarius*
  Warbling Vireo, *Vireo gilvus*
  Philadelphia Vireo, *Vireo philadelphicus*
  Red-eyed Vireo, *Vireo olivaceus*
  White-eyed Vireo, *Vireo griseus*
  Yellow-green Vireo, *Vireo flavoviridus*

Family Corvidae: Jays, Crows, etc. (4 semi-migratory spp.)
   Blue Jay, *Cyanocitta cristata*
   Black-billed Magpie, *Cyanocitta stelleri*
   American Crow, *Corvus brachyrhynchus*
   Chihuahua Raven, *Corvus cryptoleucus*

Family Alaudidae: Larks (1 migratory spp.)
   Horned Lark, *Eremophila alpestris*

Family Hirundinidae: Swallows and Martins (8 migratory spp.)
   Purple Martin, *Progne subis*
   Tree Swallow, *Tachycineta bicolor*
   Violet-green Swallow, *Tachycineta thalassina*
   Northern Rough-winged Swallow, *Stelgidopteryx serripennis*
   Bank Swallow, *Riparia riparia*
   Cliff Swallow, *Petrochelidon pyrrhonota*
   Cave Swallow, *Petrochelidon fulva*
   Barn Swallow, *Hirundo rustica*

Family Paridae: Chickadees and Tits (2 semi-migratory spp.)
   Carolina Chickadee, *Parus carolinensis*
   Black-capped Chickadee, *Parus atricapilla*

Family Sittidae: Nuthatches (2 migratory sp.)
   Red-breasted Nuthatch, *Sitta canadensis*
   White-breasted Nuthatch, *Sitta carolinensis*

Family Certhiidae: Creepers (1 migratory sp.)
   Brown Creeper, *Certhia americana*

Family Troglodytidae: Wrens (5 spp.)
   Rock Wren, *Salpinctes obsoletus*
   Carolina Wren, *Thyrothorus ludovicanus*
   House Wren, *Troglodytes aedon*
   Winter Wren, *Troglodytes troglodytes*
   Sedge Wren, *Cistothorus platensis*
   Marsh Wren, *Cistothorus palustris*

Family Polioptilidae: Gnatcatchers (2 migratory sp.)
   Blue-gray Gnatcatcher, *Polioptila caerulea*

Family Regulidae: Kinglets (2 migratory spp.)
    Golden-crowned Kinglet, *Regulus satrapa*
    Ruby-crowned Kinglet, *Regulus calendula*

Family Turdidae: Thrushes and Allies (10 migratory spp.)
    Eastern Bluebird, *Sialia sialis*
    Mountain Bluebird, *Sialia currucoides*
    Western Bluebird, *Sialis mexicana*
    Townsend's Solitaire, *Myadestes townsendi*
    Veery, *Catharus fuscescens*
    Gray-cheeked Thrush, *Catharus minimus*
    Swainson's Thrush, *Catharus ustulatus*
    Hermit Thrush, *Catharus guttatus*
    Wood Thrush, *Hylocichla mustelina*
    American Robin, *Turdus migratorius*

Family Mimidae: Mockingbirds, Thrashers and Catbirds (4 migratory spp.)
    Gray Catbird, *Dumetella carolinensis*
    Northern Mockingbird, *Mimus polyglottos*
    Sage Thrasher, *Oreoscoptes montanus*
    Brown Thrasher, *Toxostoma rufum*

Family Sturnidae: Starlings (1 migratory sp.)
    European Starling, *Sturnus vulgaris*

Family Motacillidae: Pipits (2 migratory spp.)
    American Pipit, *Anthus rubescens*
    Sprague's Pipit, *Anthus spragueii*

Family Bombycillidae: Waxwings (2 migratory spp.)
    Bohemian Waxwing, *Bombycilla garrulus*
    Cedar Waxwing, *Bombycilla cedrorum*

Family Calcariidae: Longspurs and Snow Buntings (5 migratory spp.)
    Lapland Longspur, *Calcarius lapponicus*
    Chestnut-collared Longspur, *Calcarius ornatus*
    Smith's Longspur, *Calcarius pictus*
    McCown's Longspur, *Calcarius mccownii*
    Snow Bunting, *Plectrophenax nivalis*

Family Parulidae: Wood Warblers (47 migratory spp.)
    Ovenbird, *Seiurus aurocapillus*

Worm-eating Warbler, *Helmintheros vermivorum*
Louisiana Waterthrush, *Parksia motacilla*
Northern Waterthrush, *Parksia noveboracensis*
Golden-winged Warbler, *Vermivora chrysoptera*
Blue-winged Warbler, *Vermivora pinus*
Black-and-White Warbler, *Mniotilta varia*
Prothonotary Warbler, *Protonotaria citrea*
Swainson's Warbler, *Lymnothlypis swainsonii*
Tennessee Warbler, *Oreothlypis peregrina*
Orange-crowned Warbler, *Oreothlypis celata*
Colima Warbler, *Oreothlypis crissalis*
Lucy's Warbler: *Oreothlypis luciae*
Nashville Warbler, *Oreothlypis ruficapilla*
Virginia's Warbler, *Oreothlypis virginiae*
Connecticut Warbler, *Oporornis agilis*
MacGillivray's Warbler, *Geothlypis tolmiei*
Mourning Warbler, *Geothlypis philadelphia*
Kentucky Warbler, *Geothlypis formosa*
Common Yellowthroat, *Geothlypis trichas*
Hooded Warbler, *Setophaga citrina*
American Redstart, *Setophaga ruticilla*
Kirtland's Warbler, *Setophaga kirtlandii*
Cape May Warbler, *Setophaga tigrina*
Cerulean Warbler, *Setophaga cerulea*
Northern Parula, *Setophaga americana*
Magnolia Warbler, *Setophaga magnolia*
Bay-breasted Warbler, *Setophaga castanea*
Blackburnian Warbler, *Setophaga fusca*
Yellow Warbler, *Setophaga petechia*
Chestnut-sided Warbler, *Setophaga pensylvanica*
Blackpoll Warbler, *Setophaga striata*
Black-throated Blue Warbler, *Setophaga caerulescens.*
Palm Warbler, *Setophaga palmarum*
Pine Warbler, *Setophaga pinus*
Yellow-rumped Warbler, *Setophaga coronata*
Yellow-throated Warbler, *Setophaga dominica*
Prairie Warbler, *Setophaga discolor*
Grace's Warbler, *Setophaga graciae*
Black-throated Gray Warbler, *Setophaga nigrescens*
Golden-cheeked Warbler, *Setophaga chrysoparia*
Townsend's Warbler, *Setophaga townsendi*
Black-throated Green Warbler, *Setophaga virens*

Canada Warbler, *Cardellina canadensis*
Wilson's Warbler, *Cardellina pusilla*
Red-faced Warbler, *Cardellina rubrifrons*
Painted Redstart, *Myioborus pictus*
Yellow-breasted Chat, *Icteria virens*

Family Emberizidae: Towhees and Sparrows (27 migratory spp.)
Spotted Towhee, *Pipilo maculatus*
Eastern Towhee, *Pipilo erythrophthalmus*
Green-tailed Towhee, *Pipilo chlorurus*
Cassin's Sparrow, *Aimophila cassinii*
Bachman's Sparrow, *Aimophila aestivalis*
American Tree Sparrow, *Spizella arborea*
Chipping Sparrow, *Spizella passerinea*
Clay-colored Sparrow, *Spizella pallida*
Brewer's Sparrow, *Spizella breweri*
Field Sparrow, *Spizella pusilla*
Sage Sparrow, *Artemisiospiza belli*
Vesper Sparrow, *Pooecetes gramineus*
Lark Sparrow, *Chondestes grammacus*
Lark Bunting, *Calamospiza melanocorys*
Savannah Sparrow, *Passerculus sandwichensis*
Grasshopper Sparrow, *Ammodramus savannarum*
Baird's Sparrow, *Ammodramus bairdii*
Henslow's Sparrow, *Ammodramus henslowii*
Le Conte's Sparrow, *Ammodramus leconteii*
Nelson's Sparrow, *Ammodramus nelsoni*
Fox Sparrow, *Passerella iliaca*
Song Sparrow, *Melospiza melodia*
Lincoln's Sparrow, *Melospiza lincolnii*
Swamp Sparrow, *Melospiza georgiana*
White-throated Sparrow, *Zonotrichia albicollis*
Harris's Sparrow, *Zonotrichia querula*
White-crowned Sparrow, *Zonotrichia leucophrys*
Dark-eyed Junco, *Junco hyemalis*

Family Cardinalidae: Cardinals, Tanagers and Grosbeaks (12 migratory spp.)
Summer Tanager, *Piranga rubra*
Scarlet Tanager, *Piranga olivacea*
Western Tanager, *Piranga ludoviciana*
Hepatic Tanager, *Piranga flava*
Rose-breasted Grosbeak, *Pheucticus ludovicianus*

Black-headed Grosbeak, *Pheucticus melanocephalus*
Blue Grosbeak, *Passerina caerulea*
Lazuli Bunting, *Passerina amoena*
Indigo Bunting, *Passerina cyanea*
Varied Bunting, *Passerina versicolor*
Painted Bunting, *Passerina ciris*
Dickcissel, *Spiza americana*

Family Icteridae: Blackbirds and Orioles (16 migratory spp.)
Bobolink, *Dolichonyx oryzivorus*
Red-winged Blackbird, *Agelaius phoeniceus*
Eastern Meadowlark, *Sturnella magna*
Western Meadowlark, *Sturnella neglecta*
Yellow-headed Blackbird, *Xanthocephalus xanthocephalus*
Rusty Blackbird, *Euphagus carolinus*
Brewer's Blackbird, *Euphagus cyanocephalus*
Common Grackle, *Quiscalus quiscula*
Great-tailed Grackle, *Quiscalus mexicanus*
Bronzed Cowbird, *Molothrus aeneus*
Brown-headed Cowbird, *Molothrus ater*
Orchard Oriole, *Icterus spurius*
Bullock's Oriole, *Icterus bullockii*
Baltimore Oriole, *Icterus galbula*
Hooded Oriole, *Icterus cucullatus*
Scott's Oriole, *Icterus parisorum*

Family Fringillidae: Finches (12 migratory or semi-migratory spp.)
Gray-crowned Rosy-Finch, *Leucosticte tephrocotis*
Pine Grosbeak, *Pinicola enucleator*
Purple Finch, *Haemorhous purpureus*
Cassin's Finch, *Haemorhous cassinii*
House Finch, *Haemorhous mexicanus*
Red Crossbill, *Loxia curvirostra*
White-winged Crossbill, *Loxia leucoptera*
Common Redpoll, *Acanthus flammea*
Pine Siskin, *Spinus pinus*
Lesser Goldfinch, *Carduelis psaltria*
American Goldfinch, *Carduelis tristis*
Evening Grosbeak, *Coccothraustes vespertinus*

Literature Cited &
Selective Bibliography

Able, K. P. (ed.) 1999. *Gathering of Angels: Migrating Birds and their Ecology.* Ithaca, NY: Cornell Univ. Press.

Alerstam, T., and D. A. Christie. 2012. *Bird Migration.* Cambridge, UK: Cambridge Univ. Press.

Baird, J. 1999. Returning to the tropics: The epic flight of the Blackpoll Warbler. Pp. 63-77, in *Gathering of Angels: Migrating Birds and their Ecology* (K. P. Able, ed.). Ithaca, NY: Cornell Univ. Press.

Baumgartner, F., and A. M. Baumgartner, 1993. *Oklahoma Bird Life.* Norman, OK: Univ. of Oklahoma Press.

Benkman, C. W. 1990. Adaptation to single resources and the evolution of crossbill (*Loxia*) diversity. *Ecol. Mongr.* 62=3:305-325.

Berthold, P. 2012. *Bird Migration: A General Survey.* New York, NY: Oxford Univ. Press.

Bildstein, K. L. 1969. Racing with the sun: The forced migration of the Broadwinged Hawk. Pp. 79-102, in *Gathering of Angels: Migrating Birds and their Ecology* (K. P. Able, ed.). Ithaca, NY: Cornell Univ. Press.

Brown, C. R., and M. B. Brown. 1996. *Coloniality in the Cliff Swallow: The Effect of Group Size on Colony Behavior.* Chicago: Univ. of Chicago Press.

Brown, M., S. Dinsmore, and C. R. Brown. 2012. *The Birds of Southwestern Nebraska.* Lincoln, NE: Division of Conservation and Survey, Univ. of Nebraska School of Natural Resources.

Brown, M., and P. A. Johnsgard. 2012. *Birds of the Central Platte Valley and Adjacent Counties.* Lincoln, NE: Zea Books, in preparation.

Buckley, P. A., M. S. Foster, E. S. Morton, R. S. Ridgley, & F. G. Buckley (eds.) 1985. *Neotropical Ornithology.* Ornith. Monogr. 36. Washington, D.C.: American Ornithologists' Union.

Cable, T. T., S. Seltman and K. J. Cook. 1986. *Birds of Cimarron National Grassland.* Ft. Collins: U.S.D.A., Forest Service Gen. Tec. Rep. RM-GTR-281, Rocky Mtn. Forest & Range Experiment Station. 108 pp.

Calder, W. A. 1999, Hummingbirds in Rocky Mountain meadows. Pp.149-168, in *Gathering of Angels: Migrating Birds and their Ecology* (K. P. Able, ed.). Ithaca, NY: Cornell Univ. Press.

Canterbury, J., & P. A. Johnsgard. In prep. *Birds and Birding in the Bighorn Mountains Region, Wyoming.*

Cox, C. W. 2010. *Bird Migration and Global Change.* Washington, D.C.: Island Press.

Chesser, R. T. 1994. Migration in South America: An overview of the austral system. *Bird Conservation International* 4:91–107.

Crump, D. J. (ed.) 1984. *A Guide to our Federal Lands.* National Geographic Soc., Washington, D.C.

Curson, J., D. Quinn and D. Beadle. 1994. *Warblers of the Americas.* Boston: Houghton Mifflin.

DeGraaf, R. M., and J. H. Rappole. 1995. *Neotropical Migratory Birds: Natural History, Distribution, and Population Change.* Cornell University Press. Ithaca, New York. 676 pages.

Dinsmore, S. J., L. S. Jackson, B. L. Ehrsman, and J. J. Dinsmore. 1995. *Iowa Wildlife Viewing Guide.* Falcon Press, Helena, MT,

Eaton, S. W. 1953. Wood warblers wintering in Cuba. *Wilson Bull.* 65: 169–74.

Elphick. J. 1995. *Collins Atlas of Bird Migration.* London, UK: HarperCollins.

Farrar, J, 2004. Birding Nebraska. 2004. *Nebraskaland Magazine,* 82(1), pp. 1–178.

Finch, D. M. and P. W. Stangel. 1992. *Status and Management of Neotropical Migratory Birds.* USDA Forest Service, General Technical Report RM-229. 422 pp.

Finch, D. M., and T. Martin. 1991. *Research Working Group of the Neotropical Migratory Bird Program: Workplans and Reports,* 18 October 1991. U.S. Dept. Agric., Forest Serv., Rocky Mountain Forest and Range Exp. Sta., Laramie, Wyoming. 422 pp.,

Fitzpatrick, J. W, 1980, Wintering of North American tyrant flycatchers in the Neotropics. Pp. 67–78, in *Migrant Birds in the Neotropics: Ecology, Behavior, Distribution, and Conservation* (A. Keast and E. S. Morton, eds.). Washington, D. C.: Smithsonian Inst. Press, 576 pp.

Fuller, M. R., W.S. Seegar and L. S. Schueck, 1988. Routes and travel rates of migrating Peregrine Falcons *Falco peregrinus* and Swainson's Hawks *Buteo swainsoni* in the western hemisphere. *J. Avian Biol.* 20: 433-440.

Gauthreaux, S. A. 1971. Radar and direct visual study of passerine spring migration in southern Louisiana. *Auk* 88: 487-500.

Gauthreaux, S. A. 1999, Neotropic migrants and the Gulf of Mexico: The view from above. Pp. 51-62, in *Gathering of Angels: Migrating Birds and their Ecology* (K. P. Able (ed.). Ithaca, NY: Cornell Univ. Press.

Gill, K., & P. A. Johnsgard. 2010. The Whooping Cranes: Survivors against all odds. *Prairie Fire,* Sept., 2010, pp. 12, 13, 16, 22.

Greenberg, R., and P. P. Marra (eds.) 2005. *Birds of Two Worlds: The Ecology and Evolution of Migration.* Baltimore, MD: Johns Hopkins Univ. Press.

Gress, Bob, and P. Jensen, 2008. *Kansas Birds and Birding Hot Spots.* Lawrence, KS: Univ. Press of Kansas.

———, and G. Potts. 1993. *Watching Kansas Wildlife: A Guide to 101 Sites.* Lawrence: Univ. Press of Kansas.

Hagan, J. M., III, and D. W. Johnston, (eds.). 1992. *Ecology and Conservation of Neotropical Migrant Landbirds.* Washington, D.C.: Smithsonian Inst. Press.

Haines, A. M., M. J. McGrady, M. S. Martell, B. J. Dayton, M. B. Henke and W. S. Seegar. 2003. Migration routes and wintering locations of Broad-winged Hawks tracked by radio-telemetry. *Wilson Bull.* 115: 166-169.

Harrington, B. A., and R. I. G. Morrison. 1979. Semipalmated Sandpiper migration in North America. *Studies in Avian Biology* 2L 83–100.

Harrington, B. A., F. J. Leeuwenberg, S. L. Resende, and R. McNeil. 1991. Migration and mass change of White-rumped Sandpipers in North and South America. *Wilson Bull.* 103:621-636.

Harris, G. 1998. *A Guide to the Birds and Mammals of Coastal Patagonia.* Princeton, NJ: Princeton Univ. Press.103: 621-636

Hayman, P., J. Marchant and T. Prater. 1986. *Shorebirds: An Identification Guide.* Boston, MA: Houghton Mifflin.

Henderson, C. L., et al. 1997. *Traveler's Guide to Wildlife in Minnesota.* St. Paul, MN: Minnesota Dept. of Natural Resources.

Hjertaas, D. G., D. H. Ellis, D. W. Johns, and S. L. Moon. 2001. Tracking Sandhill Crane migration from Saskatchewan to the Gulf coast. *Proc. N. Am Crane Workshop* 8:57-61.

Hochbaum, H. A. 1944. *The Canvasback on a Prairie Marsh.* Washington, D.C: Wildlife Management Institute, and Harrisburg, PA: Stackpole.

Hochbaum, H. A. 1967. *Travels and Traditions of Waterfowl.* Minneapolis, MN: University of Minnesota Press.

Jehl, J. R., Jr. 1979. The autumnal migration of the Baird's Sandpiper. *Studies in Avian Biol.* 2:55–68.

Johnsgard. P. A. 1979. *Birds of the Great Plains: Breeding Species and their Distribution.* Revised ed. with a Literature Supplement and revised maps. 2009. http://digitalcommons.unl.edu/bioscibirdsgreatplains/1/

———. 1980. Where have all the curlews gone? *Natural History,* August, p. 30-34.

———. 1995. *This Fragile Land: A Natural History of the Nebraska Sandhills.* Lincoln, NE: University of Nebraska Press.

———. 2007. *The Birds of Nebraska.* Lincoln, NE: Published by author. Revised edition . http://digitalcommons.unl.edu/biosciornithology/38

————. 2008. *A Guide to the Natural History of the Central Platte Valley of Nebraska.* http://digitalcommons.unl.edu/biosciornithology/40

————. 2010. *Waterfowl of North America.* Revised, with a 2009 Literature Supplement. http://digitalcommons.unl.edu/biosciwaterfowlna/

————. 2012a. *Wetland Birds of the Central Plains: South Dakota, Nebraska and Kansas.* Lincoln, NE: http://digitalcommons.unl.edu/zeabook/8

————. 2012b. *The Nebraska Wetlands and their Ecology.* Lincoln, NE: Division of Conservation and Survey, Univ. of Nebraska School of Natural Resources.

Johnsgard, P. A. and T. S. Shane, 2009. Four Decades of Christmas Bird Counts in the Great Plains: Ornithological Evidence of a Changing Climate. 2009, URL: http://digitalcommons.unl.edu/biosciornithology/46/

Jones, J. O. 1990. *Where the Birds Are: A Guide to all 50 States and Canada.* New York, NY: Wm. Morrow.

Jorgensen, J. 2012. *Birds of the Rainwater Basin, Nebraska.* Nebraska Game & Parks Dept. website: http://outdoornebraska.ne.gov/../NGBirds/../Birds%20of%20the%20Rainwater%20Basin%20Version%201

Kear, J,. (ed.) 2005. *Ducks Geese and Swans.* 2 vols. Oxford, UK: Oxford Univ. Press.

Keast, A. 1980. Spatial relationships between migratory parulid warblers and their ecological counterparts in the Neotropics. Pp. 109–30, in *Migrant Birds in the Neotropics: Ecology, Behavior, Distribution, and Conservation* (A. Keast and E. S. Morton, eds.). Washington, D. C.: Smithsonian Inst. Press, 576 pp.

Keast, A., and E. S. Morton. 1980. *Migrant Birds in the Neotropics: Ecology, Behavior, Distribution, and Conservation.* Washington, D. C.: Smithsonian Inst. Press, 576 pp. QL 687- AiM53.

Kingery, H. E. 2007. *Birding Colorado.* Helena, MT: Falcon Guides.

Krapu, G. 1999. Sandhill cranes and the Platte River. Pp. 103-117, in: *Gathering of Angels: Migrating Birds and their Ecology* (K. Able, ed.). Ithaca, NY: Cornell Univ. Press.

Krapu, G., G. C. Iverson, K. J. Reinecke and C. M. Boise. 1985. Fat deposition and usage by arctic-nesting Sandhill Cranes during spring. *Auk* 102:362-368.

Kutac, E. A. 1998, *Birder's Guide to Texas.* 2nd ed. Houston, TX : Ed. Gulf Publ. Co.

Kuyt, E. 1992. Aerial radio-tracking of Whooping Cranes migrating between Wood Buffalo National Park and Aransas National Wildlife Refuge. *Canadian Wildlife Service Occasional Papers* 7:1-53.

Lincoln, F. C. 1935. *The Waterfowl Flyways of North America.* Washington, D.C.: USDA Circ. 342. 12 pp.

————. 1943, Migration Routes and Flyways. Pp. 47–53, in *Ducks, Geese and Swans of North America*, F. H. Kortright. 1st. ed. Washington D. C.: American Wildlife Institute.

Linduska, J. P. (ed.) 1964. *Waterfowl Tomorrow*. Washington, D.C. U. S. Dept. of Interior., Bureau of Sport Fisheries and Wildlife.

Lingle, G. 1994. *Birding Crane River*. Grand Island, NE: Harrier Publ. Co.

Lowery, G. H., Jr. 1945. Evidence for trans-Gulf migration and the coastal hiatus. *Wilson Bull.* 57:92123.

Madge, S. and H. Burn. 1988. *Wildfowl*. London, UK: Christopher Helm.

Martin, T. E., and D. M. Finch (eds.). 1995. *Ecology and Management of Neotropical Migratory Birds*, New York, NY: Oxford Univ. Press.

McClintock, E. P., T. C. Williams, and T. M. Teal. 1978. Autumn bird migration observed from ships in the western North Atlantic. *Bird-Banding* 49:262272,

McEneaney, T. 1993. *The Birder's Guide to Montana*. Helena, MT: Falcon Press.

Moorman, T. E., and P. N. Gray. 1994. Mottled Duck. *The Birds of North America*, No. 81. Philadelphia, PA: The Academy of Natural Sciences, and Washington, D.C.: American Ornithologists' Union.

Morrison R. I. G., and R. K. Ross. 1989. *Atlas of Nearctic Shorebirds on the Coast of South America*. Can. Wildlife Serv. Spec. Publ., vol. 1

Myers J. P. 1983. Conservation of migrating shorebirds; Staging areas, geographic bottlenecks and regional movements. *Am. Birds* 37:23–35.

Myers J. P., And L. P, Myers. 1979. Shorebirds of coastal Buenos Aires Province, Argentina. *Ibis* 121:186–200.

Newton, I. 2007. *The Migration Ecology of Birds*. New York, NY: Academic Press.

————. 2010. *Bird Migration*. New Naturalist 113. London, UK: Collins.

Nisbet, I, C, T, 1970. Autumn migration of the Blackpoll Warbler: Evidence for long flight provided by regional survey. *Bird-Banding* 41: 207240.

Pena, M. de la, and M. Rumbull. 2001. *Birds of Southern South America and Antarctica*. Princeton, NJ: Princeton Univ. Press.

Perrins, C. M., and J. Elphick. 2001. *The Complete Encyclopedia of Birds and Bird Migration*. New York. NY: Harper-Collins.

Pettingill, O, S., and N. R. Whitney. 1965. *Birds of the Black Hills*. Special Publication No. 1, Ithaca, NY. Ithaca, NY: Cornell Laboratory of Ornithology.

Ralph, C J. 1978. The disorientation and possible fate of young passerine coastal migrants. *Bird-Banding* 49:237247.

Rappole, J. H., E. S. Morton, T. E. Lovejoy III and J. L. Ruos. 1983. *Nearctic Avian Migrants in the Neotropics*. Washington, D.C.: U.S. Fish & Wildlife Service.

Ridgely, R. S., and G. Tudor. 1989. *The Birds of South America. The Oscine Passerines.* Vol. 1. Austin, TX: Univ. of Texas Press.

———. 1994. *The Birds of South America. The Suboscine Passerines.* Vol. 2. Austin, TX: Univ. of Texas Press.

Robinson, S. K., J, W. Fitzpatrick and J. Terborgh. 1995. Distribution and abundance of Neotropical migrant land birds in the Amazon basin and Andes. *Bird Conserv. Internat.* 56;305–323

Sauer, J. R., J. E. Hines, J. E. Fallon, K. L. Pardieck, D. J. Ziolkowski, Jr., and W. A. Link. 2011. *The North American Breeding Bird Survey, Results and Analysis 1966 - 2010. Version 12.07.2011* USGS Patuxent Wildlife Research Center, Laurel, MD.

Scharf, W C., J. Kren, L. R. Brown and P. A. Johnsgard. 2008. Body Weights and Distributions of Birds in Nebraska's Central and Western Platte Valley. U. of Nebraska Digital Commons, *Papers in Ornithology*: http://digitalcommons.unl.edu/biosciornithology/43

Scott. O. K. 1993. A *Birder's Guide to Wyoming.* Colorado Springs, CO: American Birding Association.

Senner, S. E., and E. F. Martinez. 1982. A review of Western Sandpiper migration in interior North America. *Southwestern Nat.* 27: 149–159.

Shepard, L. 1996. *The Northern Plains. The Smithsonian Guides to Natural America.* Washington, D.C.: Smithsonian Books.

Skagen, S. K., and F. L. Knopf. 1993. Toward conservation of mid-continental shorebird migrations. *Conserv. Biol.* 7:533–41.

Stinson, C. H. 1977.The spatial distribution of wintering Black-bellied Plovers. *Wilson Bull.,* 89:470–2.

Stotz, D. F., R. O. Bieregaard, M. Cohn-Halt, P. Petermann, J. Smith, A. Whittaker, and S. V. Summer. 1992. The status of North American migrants in central Brazil. *Condor,* 94: 608–21.

Stotz, D. F, J. W. Fitzpatrick, T. A. Parker, III, & D. K. Moskovits. 1996. *Neotropical Birds: Ecology and Conservation.* Chicago: University of Chicago Press. 477 pp.

Tallman, D. A., D. L. Swanson & J. S. Palmer, 2002. *Birds of South Dakota.* Aberdeen, SD: South Dakota Ornithologists' Union.

Terborgh, J. W. 1980. The conservation status of Neotropical migrants. Pp., 21–30, in *Migrant Birds in the Neotropics: Ecology, Behavior, Distribution, and Conservation* (A. Keast and E. S. Morton, eds.). Washington, D. C.: Smithsonian Inst. Press, 576 pp.

Terborgh, J. W. 1980. Where *have all the Birds Gone? Essays on the Biology and Conservation of Birds that Migrate to the Neotropics.* Princeton, NJ: Princeton Univ. Press.

Thompson, M. C., C. A. Ely, B. Gress, C. Otte, S. T. Patti, D. Seibel, and E. A. Young. 20112. *Birds of Kansas.* Lawrence: Univ. Press of Kansas.

Tramer, E. J., and T. R. Kemp. 198, Notes on migrants wintering at Monteverde, Costa Rica. *Wilson Bull.,* 94: 350354.

U. S. Fish & Wildlife Service, 2009. *2009 Waterfowl Population Status.* Washington, D.C., Administrative Report. U. S. Dept. of Interior.

van Perlo, B. 2005. *Birds of Central & South America.* Princeton, NJ: Princeton Univ. Press.

Wetlands International. 2002. *Waterfowl Population Estimates.* 3rd. ed. Wegeninen, the Netherlands. Wetlands International Global Series, No. 12.

Wetmore, A. 1965–1984. *Birds of the Republic of Panama.* 4 vols. Washington, D. C.: Smithsonian Inst. Press.

Williams, T. C., J. M. Williams, L. C. Ireland, and J. M. Teal. 1977. Autumnal bird migration over the western North Atlantic Ocean. *Am. Birds* 3:252-267.

Wishart, D. (ed.) 2004. *Encyclopedia of the Great Plains.* Lincoln. NR: Univ. of Nebraska Press.

Young, M. T. 2000. *Colorado Wildlife Viewing Guide.* 2nd ed. Falcon Press, Helena, MT.

Zimmer, K J. 1979. A *Birder's Guide to North Dakota.* Denver, CO: L & P. Press,

Zimmerman, J. L. 1990. *Cheyenne Bottoms: Wetland in Jeopardy.* Lawrence, KS: Univ. Press of Kansas.

———. and S. T. Patti. 1988. *A Guide to Bird-finding in Kansas and Western Missouri.* Lawrence: Univ. Press of Kansas.

The University of Nebraska–Lincoln does not discriminate
based on gender, age, disability, race, color,
religion, marital status, veteran's status,
national or ethnic origin,
or sexual orientation.

Nebraska
UNIVERSITY OF
Lincoln®

.

www.ingramcontent.com/pod-product-compliance
Lightning Source LLC
Chambersburg PA
CBHW031505270326
41930CB00006B/258